The Art of
The STONEMASON

THE ART OF THE STONEMASON

by Ian Cramb

Alan C. Hood & Company, Inc.

CHAMBERSBURG, PENNSYLVANIA

ISBN 0-911469-27-3

Published by Alan C. Hood & Co., Inc.
Chambersburg, PA 17201

www.hoodbooks.com

Copies of *The Art of the Stonemason* may be obtained by sending $28.50 per copy to:

Alan C. Hood & Co., Inc.
P.O. Box 775, Chambersburg, PA 17201

Price includes postage and handling.
Quantity discounts are available to dealers and non-profit organizations.
Write on letterhead for details.

Cover adapted from the original design of
Rick Britton by James F. Brisson
Photographs by Ian Cramb
Illustrations by Ian Cramb

———◆◆———

Library of Congress Cataloging-in-Publication Data

Cramb, Ian 1928-
    The Art of the Stonemason / by Ian Cramb.
        p. cm.
    Reprint. Originally published: White hall, VA : Betterway Books, 1992.
    Includes bibliographical references and index.
    ISBN 0-911469-27-3
    1. Stonemasonry--Amateurs' manuals.   I. Title
    TH5411.C73 2006
    693'.1--dc22
                                            2006041167

            10 9   8   7   6   5   4   3   2

I dedicate this book to my wife Betty,
my oldest son Ewan, to try and keep this old craft going,
and my young son John, my faithful helper for so many years.

I also dedicate the book to all the old bygone stonemasons, who built the old structures, for the
knowledge I have gained through doing restoration work.

Contents

Introduction

I have written this book, *The Art of the Stonemason*, as a way of sharing my experience, as practitioner and as administrator, in building with stone and in restoration work using stone.

The methods and mortar mixes I use in stonework are traditional, as handed down through the generations. They are easily explained. The practical problems are arranged in order from the most simple to the more complex, and in their description, I have avoided technicalities as much as possible.

My efforts should be of interest and help to stonemasons, architects, and anyone interested in building and restoration work using stone.

Proficiency in any art cannot be obtained without study and perseverance. Stonework is certainly no exception, but difficulties will soon vanish if you commence with the confidence that you will succeed.

Practice makes perfect.

Ian Cramb
Building and restoration stonemason
1991

The Traditional Method of Building Random Rubble

What is *random rubble*? Random rubble is the traditional or Early Celtic art of building with stone. It uses stones that are not squared but are of irregular sizes and are bedded on mortar or other suitable bedding material. It takes time and practice to "do" random rubble in the traditional method.

Before you attempt to start and build random rubble, have a good look around at some old stone walls, such as those found on old farm houses, country churches, and boundary walls. These will give you a general idea of the style and type of stone used, which would have been quarried locally, or more likely, gathered from the fields. No drawing can give you the position of stones on the wall in rubble work. The shapes and sizes of the individual stones selected determine their positions in the wall. You must try to imagine what they will look like; this is part of the secret of doing random rubble.

Having selected a building showing a good example of random rubble, stand back and look along the wall. You will notice all random rubble walls have one thing in common; that is, they have been built in 18-inch to 2-foot *lifts* (heights), which style of building is known as *coursed random rubble*. This is the one and only way of building a random rubble wall. You should also notice that each stone runs level on its *bed* from the smallest to the largest, and each lift on the wall is dead level. To save showing straight, long lines every lift, *risers* (stones that rise above or through the line) are placed at fairly regular intervals to break the straight line.

Choosing Your Stone

Before you start building your garden wall, or stone-facing the front of your house, or doing any type of rubble work, think of the type of stone you are going to use. You would not want a cluster of large stones to build a small garden wall, for example.

I have observed some beautiful old stone walls, scarred and patched with large, rough, squared stones by inexperienced tradesmen. The original stone from the wall was uplifted and dumped. The patched area was pointed over with a three-to-one soft sand mix and rubbed with a rag. This eye-catching mess forms a very weak patch, as it is the strong against the weak. It will only hold a very short time before it collapses in a solid lump, bringing down more of the surrounding area. You have to try to rebuild, blending in with the existing stonework, using the proper mortar mixes.

The best type of stone for your building work is preferably from old rubble walls, as they already have a prepared face. You also need stone that suits the height and length of your wall. You will have to lift everything from the smallest stone to the largest you can handle. It is wasted effort to try to lift a stone that is too heavy. You may waste more time and energy by cutting it down to size. Choose the right size of stone to begin.

Having selected and gathered your stone, drop the stone about three feet from the site of your wall. This distance allows you easy access for building. Do not try to select stone for size and color. What you lift, you build; you will find the wall will work out for the appropriate stone size and color. Remember the plan and the wall you are building are all in your mind at this point. Below I explain in simple terms how to build this style of stonework. When you look at the first few square feet of completed wall, it almost never looks right. But do not let this put you off. I get this same feeling. It will look better as you progress.

Take your time, as practice makes perfect. If you make an error, rectify it before going on.

Random rubble — traditional method.

Traditional random rubble, with flat arch over window.

Remember, your work will be seen and admired for years to come, and there is no way to cover up your mistakes. As the old Scottish saying goes, "only fools and wimen" criticize half-finished stonework. Your finished work *will* be achieved. I showed a friend of mine, a plumber by trade, the rudiments of random rubble building. He built the back and sides of his house with random rubble, and I am proud to say it is beautiful and well done.

Beginning the Work

The following is an easy method for building a garden or boundary wall, in random rubble, *double-faced* (meaning the wall has both front and rear faces), not to exceed six feet in height.

For the thickness of your wall, plan on between 15 and 17 inches, depending on your stone. Fifteen inches is the minimum for a double-faced wall; 7 1/2 inches for a single-faced wall with a brick or block backing wall.

Having determined your thickness, stretch a line between two pegs to give you the front face building line. This will also give you a guide for foundation excavations.

With this foundation line, add 3 inches of extra width on either face of the proposed wall. This will form a *scarcement* (ledge) on your foundations (Figure 1).

Before you start on the foundations, put in some wooden pegs (Figure 1), leveling through from each peg. These will act as your *datum line* — the point from which you will measure up for the levels on each lift (18 to 24 inches). If your pegs are level throughout, your wall line will be level.

Start to excavate for your foundations. Remove the topsoil and dig down until you reach the dense subsoil. Dig about 9 to 12 inches below the topsoil level; if the soil is clay, dig down an extra 6 inches. Make sure your excavated foundation track is level, even if your ground slopes. On sloping ground your foundation track must have "steps," and each step in the excavated track must be level (Figure 2).

Do not use concrete for foundations. I never have. You will find all old stone buildings built directly on the ground. Any settlement that does occur will be taken up within the wall. If you use the proper mortar mixes, cracks will not show. With a solid concrete base, any movement or settlement crack affects the whole wall. This action also applies whenever you use strong cement mixes in building.

Set your stones on the mix you are using for building your wall. Start building your foundation (with scarcement) up to about 2 inches below ground level (Figure 3).

Your foundation should measure, upon completion, the thickness of the wall plus 6 inches: 3 inches on either side to form the scarcement. Assuming all is now level, you set the front building line from the pegs where you started.

Your lines are set. The next task is your *plumbings*, or *corners*. You can form the corners with large squared stones, called *rybits* or *quoins*. This is the easiest method. Or you can build up with the rubble you are using to build the wall, which is a little more difficult (Figure 4).

The plumbings are used to set the line for your wall and to keep the wall plumb. From these corners you will work your levels.

The Building Method

Having now built your plumbings, or corners, you set the line for the first lift — 18 to 24 inches, no higher — on each corner plumbing. This line must be level, as this is the line you are going to work up to (Figures 5, 6, 7, and 8).

If you think the line has a sag, build a center plumbing as shown in Figures 6 and 7. Then curl a piece of string around the line, with a small stone or a pad of mortar to keep it in position. This is called a *tingle*. You now have corner plumbings and a center plumbing (vertical), so what follows is a matter of filling in between the plumbings to begin forming the front face.

In building stonework, remember that your mortar mix should be sharp and damp enough to handle easily. This allows your stone to set up without squeezing mortar out and having it run down the face of your stonework. You set the stone on the mortar, but do not use a hammer to set it in position. When spreading the mortar on the stone, keep the mortar back an average of 1 inch from the face of the stone.

Start by setting your largest stones, say about 4 feet apart, then set the smaller stones roughly centered between the large stones. Fill in with the smaller stones overlapping each other. Avoid too many long joints (Figure 6), as it is the small stones that give the wall its strength. All stones are to be built on bed; all *snouts* (points of the stone that stick out farthest) should be built to the line. Do not use the back edge of the face of your stone as your guideline (Figure 4).

As you build up, approaching the level line, note when you have one lift to go before reaching the top of the wall. To prevent your wall from showing a straight line between each lift, allow stones to rise *through* the line every 3 or 4 feet (Figure 8). These are called *risers*, and they form a bond or tie between the lifts as you build up.

You should now be up to your front face line level, with a few risers going above the line (Figure 6). Taper as you build up your front face, allowing the back face to *tie over* (Figures 9 and 10).

Work the same method in building your back face, tying over all the time. You should finish level with the front face, with a few risers showing.

Use small stones as *infill* between the two faces, using your building mix to level off. *Do not use concrete or a strong cement mix as infill.*

Start your next lift just as described earlier. If this will be your wall head finish, or the top of the wall, build up to your line, omitting the risers (Figure 8).

Do not attempt to scrape and clean the joints immediately after building. Do not use a brush to remove excess mortar if the wall or mortar is wet, or you will end up with mortar staining, which is very difficult to remove after the wall dries.

Allow about twelve hours after you have completed a section before scraping and cleaning the joints. Do not attempt to point your wall as you would a building. Use a piece of wood, not a

metal tool, to clean the joints. Then brush downward with a stiff brush to remove all surplus mortar on the face of the wall.

Points to Remember

- All random rubble is built in courses. This is the traditional method; there is no such thing as uncoursed random rubble.

- All random rubble building must be kept level, with your lines kept level on every 18- to 24-inch lift. This breaks up the effect of having straight lines running through the wall. Put in risers (stones going above the level line) every 3 to 4 feet.

- All snouts (points of the stone that stick out farthest) must just touch the line. This means that no matter what the shape of the front face of the stone, the farthest point must just touch, if you want a straight wall. Do not use the back edge on the face of the stone as your guideline.

- A hole for every stone, and a stone for every hole. What you lift, you build.

- Never start rubble building above the height of rybits, quoins, or corner stones.

- At ground level, put in wood pegs as your level pins. These will act as your datum line. Just measure up from these points for each course. If your pins are level across, your wall line will be level. Lifts of no more than 18 to 24 inches are recommended.

- When tying the two faces of the wall together, never infill with concrete or a mass of strong cement mortar. Instead, infill with small stones and level off with the same mix with which you are building.

- The easy method for tying over is to build your front face tapered up to an 18- or 24-inch height, then do your inside face, tying over all the time.

- You can use the occasional *cant* stone, which is a stone laid on edge, or *off bed*.

- Use a semi-dry mortar mix, with sharp sand, and do not use any additives. This mix will

Traditional random rubble.

The author's neighbor, Mr. A. Nagy, built two sides and the back of his house using the traditional methods described in this book.

keep your stone in position and prevent sagging, as well as keeping the mix from running down the face of your wall.

- Never attempt to point the building as you build. Build up, point, and clean down on completion of the wall.

- Never put in a stone that is taller than it is wide (Figure 11).

- After a day's building, leave the wall to set overnight, and then scrape out the joints using a piece of wood. Then brush down the wall. Do not scrape joints or brush the wall if the mortar or wall is wet.

- Do not use concrete as a foundation for stone. Excavate to the depth of the subsoil — 7 to 9 inches if building a garden or boundary wall. Set your stones on the mix you will use for building, allowing a 3-inch projection for your scarcement, carried up to 2 inches below ground level. If the ground slopes, your *founds* (foundations) are formed in a series of steps, and you must finish level on this course between steps (see Figure 5).

- In doing your plumbings, make a chalk mark where your plumb rule or level will strike each time. Do this until you gain enough experience to "eye" down with your plumb rule. (See Figure 12, "Using Your Plumb Rule.")

- For building stone, your mortar mix should be more lime than cement. Stone and cement are not good partners. Your wall must breathe. When moisture gets into a good mortar mix, it dries out quickly. With a mix that is too strong, the moisture will stay in, and the wall will be damaged when colder temperatures come.

- You do not need a damp-proof course on a garden or boundary wall, but make sure you fit a proper cap or coping on the wall (see Figure 16 on page 33).

Building a Wall on Sloping Ground

As Figure 2 on page 20 shows, a wall built on sloping ground is built in the same random rubble method, with your stonework arranged at right angles to the pressure exerted upon them. You must line out your wall and founds, excavate your founds (forming steps), and allow 3 inches scarcement below ground level. All your steps must be level. If you do follow the line of the slope, your wall will slide. It would be impossible to have a proper end plumbing. Work in levels of 18 inches to 2 feet, using risers.

Building a Circular Wall

Figure 14 shows the method I use for building a circular random rubble wall. The tighter the circle, the smaller the stone, which keeps you from having to cut too much. Use a *trammel rod* (also known as a *beam compass*) to design a small circle. Make the trammel rod from a piece of lumber 3 inches wide and 1/2 inch thick. The length of the rod depends on the diameter of the circle. For a large circle, mark your founds with a string trammel. Work out your large circle or even a half-circle. Mark the outline on the ground, allowing for a 3-inch scarcement. This is your outside face. Now measure in for the thickness of the wall, plus the 3-inch scarcement, adjust your line to this mark, and form the inside circle.

Excavate your founds following your guidelines and build up the scarcement to about 3 inches below ground level. Having got this part all levelled out, use your trammel line and mark the outside and inside faces of the wall on top of the scarcement, forming the circle.

For small circles, use your trammel, remembering to keep the *snouts* (the points of stone that stick out farthest) to the line. Large circles (Figure 14) are built by putting your plumbings at "a" and infilling on the line of the curve. It is advisable to use your plumb rule on infillings, for both inside and outside faces. The greater the thickness of the wall, and the bigger the circle, the easier it is to build. Watch the size of stone you use; with a tighter circle you will have more cutting to do.

If you have to cut stone to follow the circle line, use the method described in the section on arch construction (Figure 34 on page 74). When building with cut dressed stone, as in stone setting, prepare your stones as for building a dressed stone arch.

Pier or Buttress with Sloping Face

Figure 15 shows how to build a pier or buttress with a sloping face. Your founds and wall are built as in random rubble, except that you cut the face of the corner stones to form the slope. Keep the snouts touching the line and the building lines level.

Coping for Random Rubble Walls

As shown in Figures 16, 17, and 18, you must level your wall head, then spread a thick bed of your mortar mix along the top of the wall. Set a line that the bottom of the rough stone must touch or be in line with. Set your stones on the mortar bed, with about 1-inch clearance between the joints. Fill this in with mortar, tapping it tight. Fill this up to about 2 to 3 inches from the top of the stone. Leave it until dry, then finish point, carrying the finished pointing to the edge of the stone wall. This will allow any water to run down the wall face instead of going down into the wall.

For a *segmental hammer finish* (Figure 16), set your stone on the mortar bed and tap down to your line, making sure the stone is bedded solidly. Allow $1/2$-inch joints, filling the joints to about $1/2$ inch from the top of the stone, and pack tightly. Scrape out the bedding joint about $3/4$ inch to allow for finishing the pointing. When you finish the pointing, press hard in both the bed joint and the coping joint to make them tight and prevent water penetration.

For the *split stone finish* (Figure 16), build up the back to the angle required, then bed your first layer of stone, allowing for a 2-inch overhang. Use a line to get your bottom stones in line. It is advisable to leave the first layer overnight to dry. For the second layer, proceed the same way. Bed on a mixture of 3 parts sand to 1 part cement and 1 part lime. Each layer of stone should be about the same length. Point all flush, except the overhanging part on each layer. This should be left clear, to avoid capillary attraction and to allow the water to drop off. When complete, paint the split stone with boiled linseed oil. This helps preserve it, gives it a nicer appearance, and makes it waterproof.

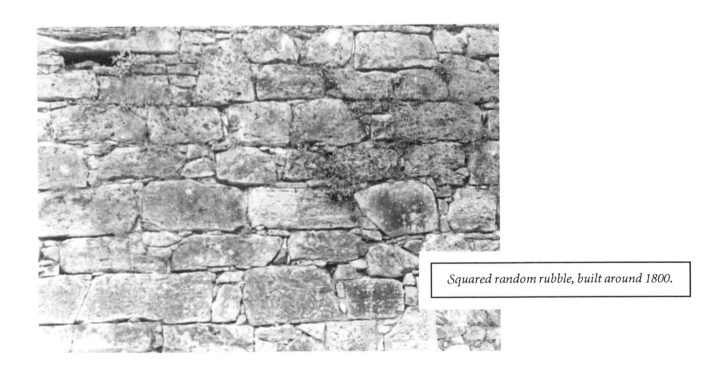

Squared random rubble, built around 1800.

Window Sills

Window sills are made in either cut stone or of a cement precast unit, with a reinforcing rod enclosed. In setting a window sill, especially when the stone or brick is resting on the stools, bed the ends only, leaving the center part hollow (*hollow bedding*). This prevents any cracking of the sill when settlement occurs. When the building is complete, you should face-point the joint with a weak mortar mix. Between the top of the sill and the timber frame, allow $\frac{1}{2}$-inch to $\frac{3}{4}$-inch clearance. Pack this with a mixture of 6 parts sand to 1 part lime and $\frac{1}{2}$ part cement, keeping it back $1\frac{1}{2}$ inches from the timber face. Once this mixture has dried out, apply linseed oil mastic, first painting all surfaces with boiled linseed oil, then pack in a sand and linseed oil mix, which should be like a stiff putty. Then clean down.

If you are using slate for your sills, you need two layers, with the top layer overlapping the joint of the bottom layer. For good bonding, paint the bottom surface with white glue and sharp sand dust, let dry, and set on a mixture of 3 parts fine sand to 1 part lime and 1 part cement. Or you can paint the bottom surface with white glue, then when it is tacky, press the slate into the mortar mix. These methods give good adhesion to thin stones. Mastic point as shown in "a" in Figure 19. Then clean down.

If you are using a slab of stone or slate for your sill, follow the process described above, using white glue on the underside for good adhesion. For added protection and to bring out the true color of your slate, rub it with boiled linseed oil until it dries in.

To make your cement-cast sill look as if it has a stone finish, dampen it, then paint it with white glue, and choose the color of sand finish you require. Put your completely dry sand in a flour sieve and sieve the sand over the wet glue. Give it a good coating of sand and leave it to dry. Then remove the excess sand. This will give you what is known as a sand-faced sill (Figure 20).

Technical Details for Dressed Copings

On all walls, whether horizontal or inclined, the coping should project on back and front, and should be a *throat* or *drip* back and front. This is a groove on the undersurface of a coping or sill that prevents water from trickling down and staining the wall, as well as making it impossible for water to get to the interior of the wall.

The top of the coping must be *weathered*; that is, it must be worked away to the edges to allow the water to get away quickly, so it will not collect and soak into the stone.

When coping is laid on a sloping wall, such as a gable, it must be supported or it will slip. A large stone with the coping worked on it to the right pitch or slope is built into the wall.

A *skew corbel* or *club skew* is a projecting stone at the lowest point of the triangular portion of the gable end of a wall. It supports the starting piece of coping and resists the sliding tendency of the coping (Figure 17).

A *kneeler* is a long stone with the coping worked on it and tailing into the gable wall. It also resists the sliding tendency of the coping (Figure 17).

Many skew corbels are constructed with a small *gablet*, which gives extra weight to the skew corbel, thus rendering it more efficient at resisting the outward thrust of the coping stones (Figure 17). The apex stones are often treated in a similar manner.

The *apex* or *saddle stone* is the highest stone of a gable. It is cut to form the termination of two adjacent inclined surfaces (Figure 17).

The *finial* is the ornament on an apex or saddle stone.

Technical Details for Dressed Stone Windows

For the following discussion, see Figure 21.

Heads are *lintols* of stone, which stretch across an opening to support the weight above.

A *transom* divides window lights horizontally. A transom may be moulded or follow the style of mullions and jambs, for which it has seatings top and bottom. The bottom edge of a transom must be throated like a sill, and for the same purpose.

Mullions are stones fixed vertically to divide a window opening into a number of lights. They rest on a seating made on the top bed of a sill for the purpose. They may be any shape to suit jambs and head.

Sills are the finish to the bottom of window openings. Their object is to throw the water away from the window and to form a covering to the wall. Sills should have at least a 2-inch projection to carry the water away from the wall. With a flush sill, the water runs down the stonework, causing staining and decay.

All overhangs in stonework, including sills, should have a groove cut underneath $5/8$ inch from the front edge and $1/2$ inch wide. This is called the *throat* or *drip*. The water gathers in this until it drops to the ground.

The *hood mould* or *label mould* is a projecting moulded ornamentation over an opening. It is useful in throwing water from the face of the wall, which would otherwise have run down windows or woodwork. All hood moulds should have a throat.

Strings are horizontal courses of stone running around a building. They are sometimes a continuation of the sills and heads; sometimes they are independent and add to the appearance, besides keeping the wall dry. They may be flush as a *band course* or *string* with the face of the wall, but are better with a projection and throated never more than 9 inches deep. Larger than this, they become *cornices*.

Mullions, transoms, and heads are secured vertically at the *seatings* or *stoolings* (the base on which the mullions, transoms, and heads rest) by means of *dowels*, which are made of copper or slate. Holes are cut in the center of the stoolings and at the ends of the mullions to receive the dowels, which are bedded in. On no account should galvanized iron or iron dowels be used, as the dampness will cause the metal to rust and burst the mullion. Dowels are pictured in Figure 67 on page 138.

Squared random rubble, built around 1800.

Figures 1 & 2

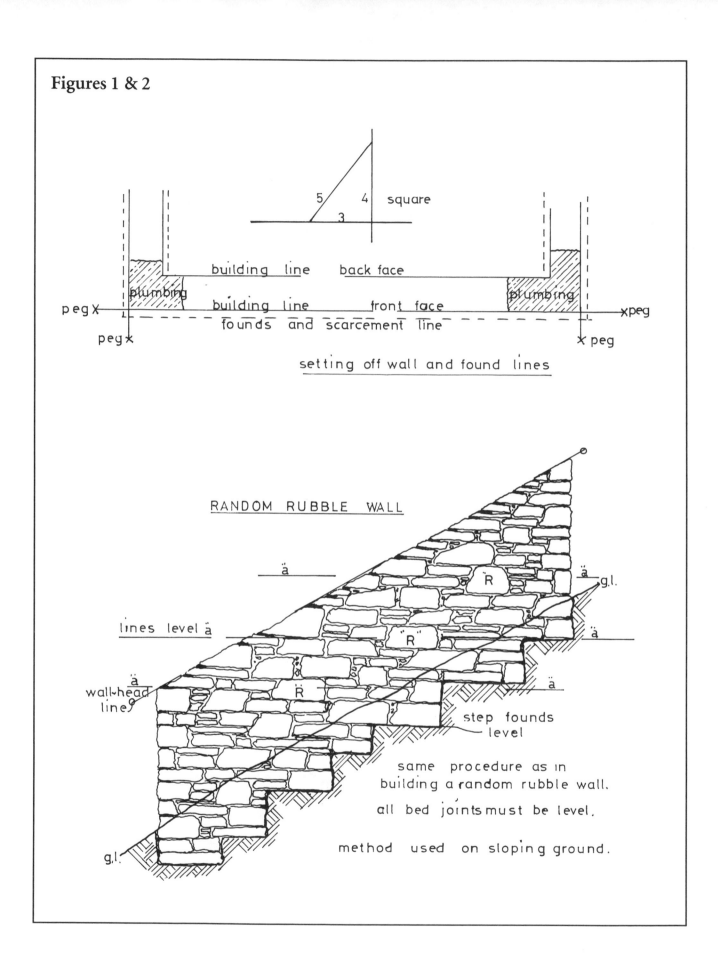

5 / 4 | square
3

building line   back face

plumbing

building line   front face

peg x   plumbing   x peg

peg x   founds and scarcement line   x peg

setting off wall and found lines

RANDOM RUBBLE WALL

ä

ä

g.l.

R

lines level ä

R

ä

ä

R

wall-head
line

step founds
level

same procedure as in
building a random rubble wall.

all bed joints must be level.

method used on sloping ground.

g.l.

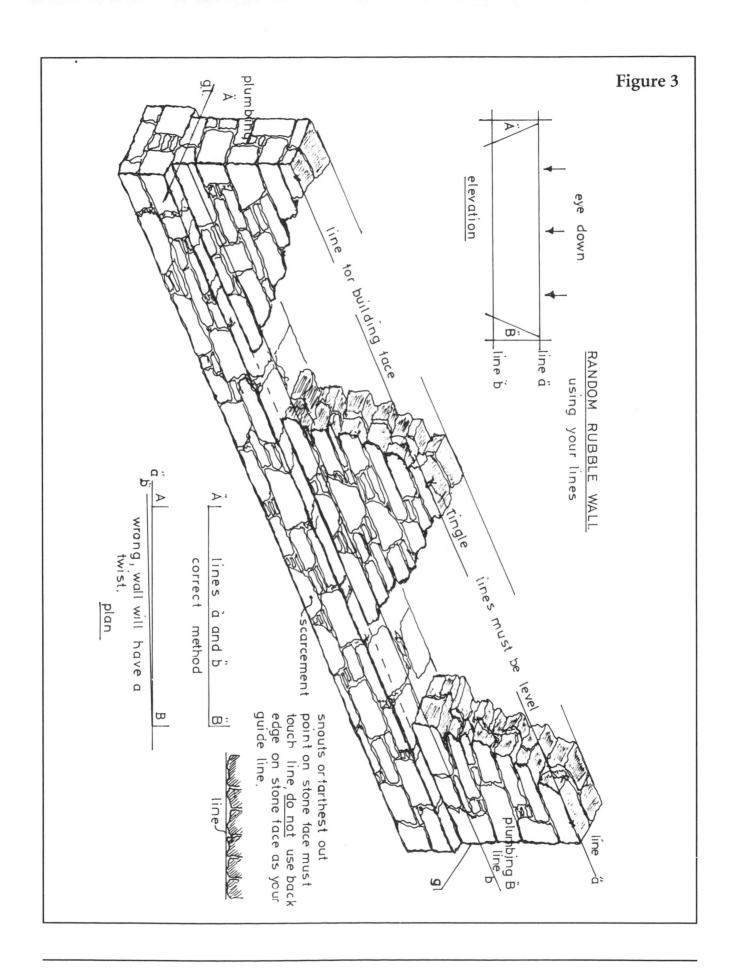

Figure 3

RANDOM RUBBLE WALL
using your lines

eye down

elevation

A
line ä
line b
B

line for building face

tingle

lines must be level

line ä

plumbing B line b

g

plumbing
Ä
g.l.

scarcement

Ä
lines ä and b
B
correct method

ä
b
A
B
wrong, wall will have a twist.

plan

snouts or farthest out point on stone face must touch line, do not use back edge on stone face as your guide line.

line

Figure 4

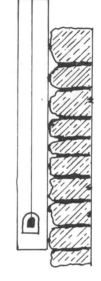

plumb rule points

corner plumbing in rubble.   using your plumb rule.

all snouts or farthest out
points of building face
of stone <u>must</u> touch
line or plumb rule

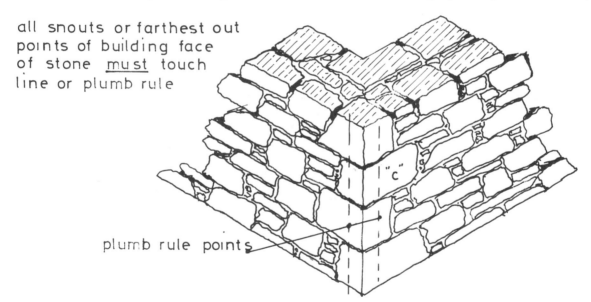

"c"

plumb rule points

corner plumbing using rough squared rybits,
or corner stones,"c".

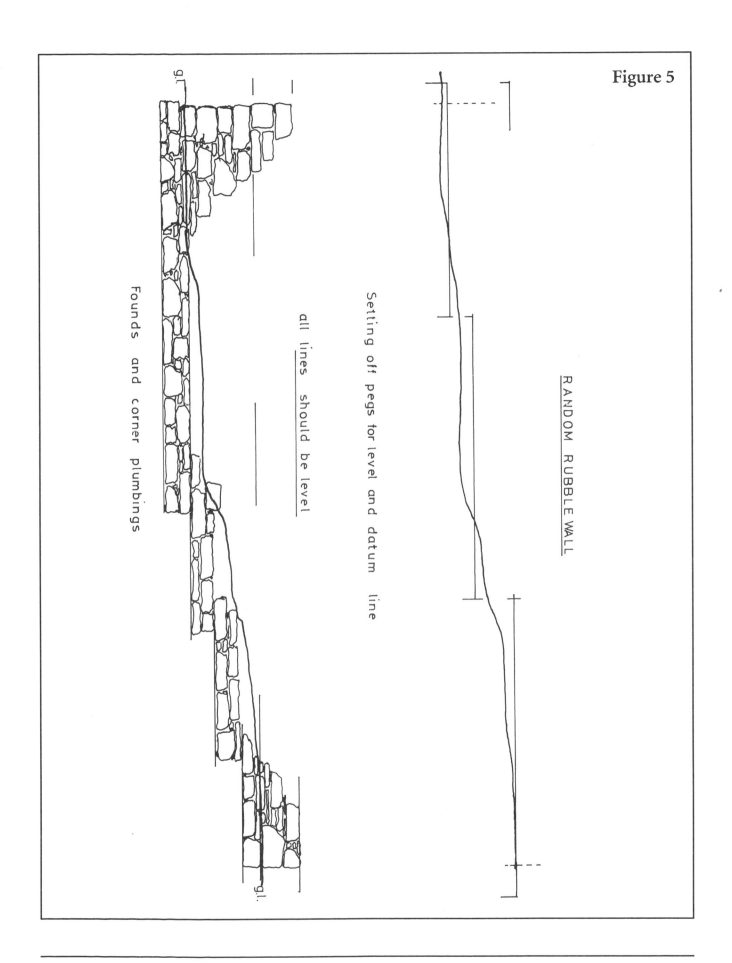

Figure 5

RANDOM RUBBLE WALL

Setting off pegs for level and datum line

all lines should be level

Founds and corner plumbings

Figure 6

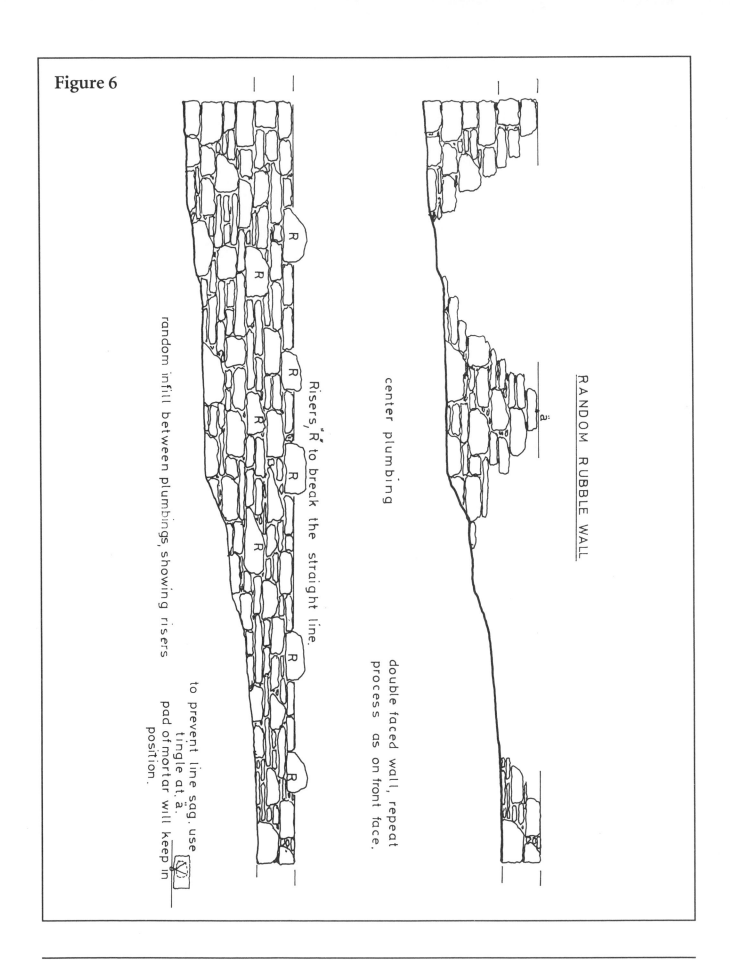

RANDOM RUBBLE WALL

center plumbing

double faced wall, repeat process as on front face.

Risers "R" to break the straight line.

random infill between plumbings, showing risers

to prevent line sag. use tingle at. ä. pad of mortar will keep in position.

Figure 7

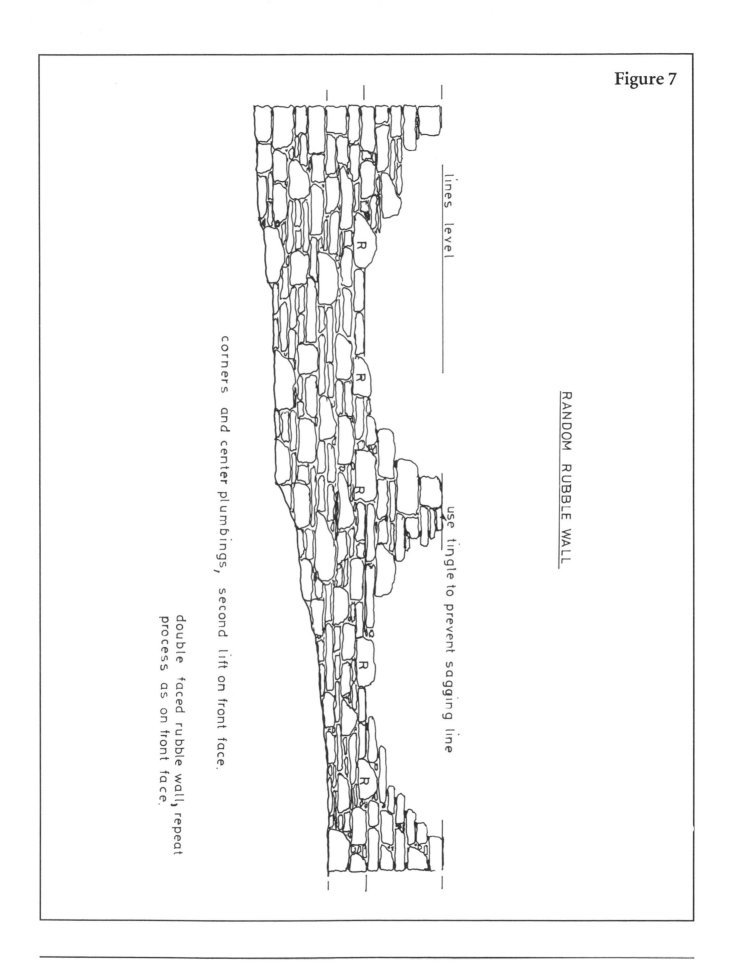

RANDOM RUBBLE WALL

lines level

use tingle to prevent sagging line

corners and center plumbings, second lift on front face.

double faced rubble wall, repeat process as on front face.

Figure 8

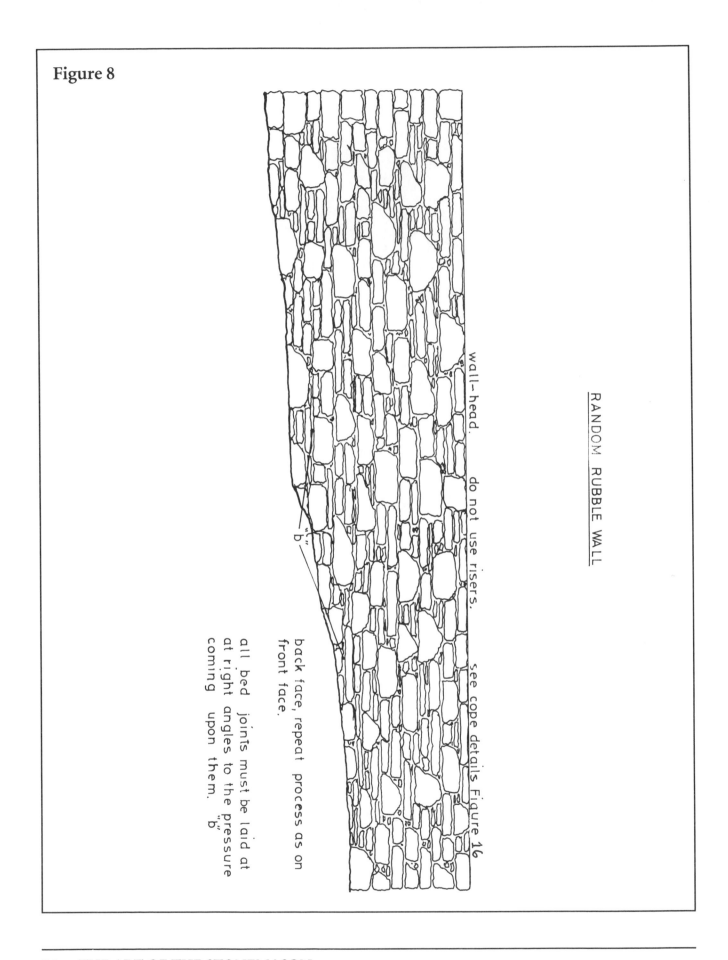

RANDOM RUBBLE WALL

wall-head.   do not use risers.   see cope details Figure 16

"b"

back face, repeat process as on front face.

all bed joints must be laid at at right angles to the pressure coming upon them. "b"

Figure 9

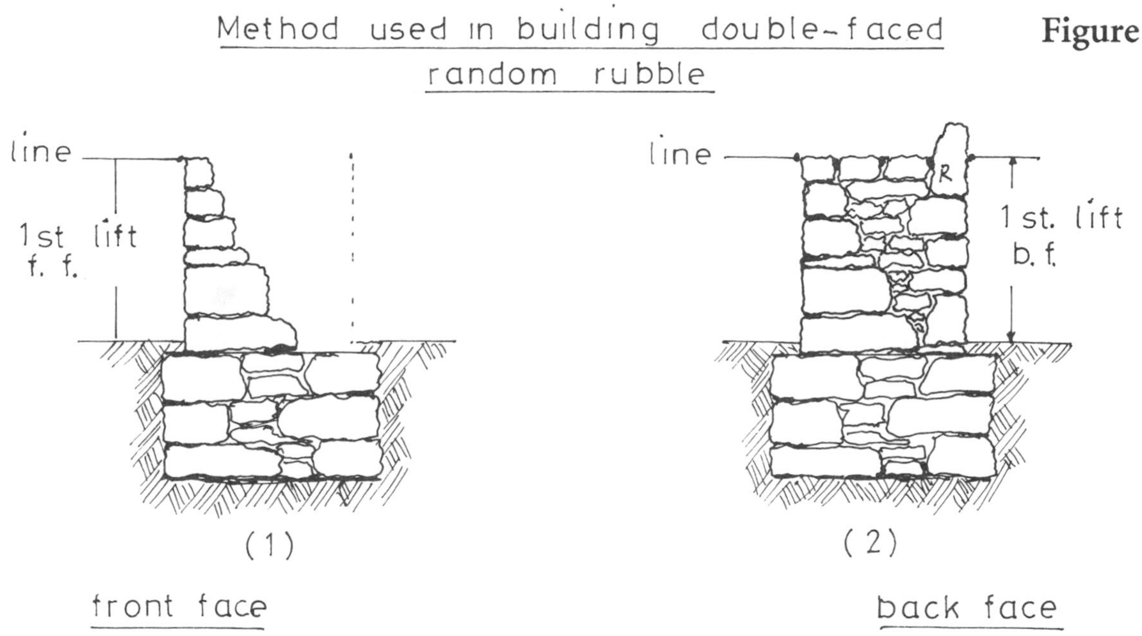

Method used in building double-faced random rubble

line

1st lift
f. f.

(1)

front face

line

1st lift
b. f.

(2)

back face

**Minimum thickness double faced wall 1' – 1".
Single faced wall 7½".**

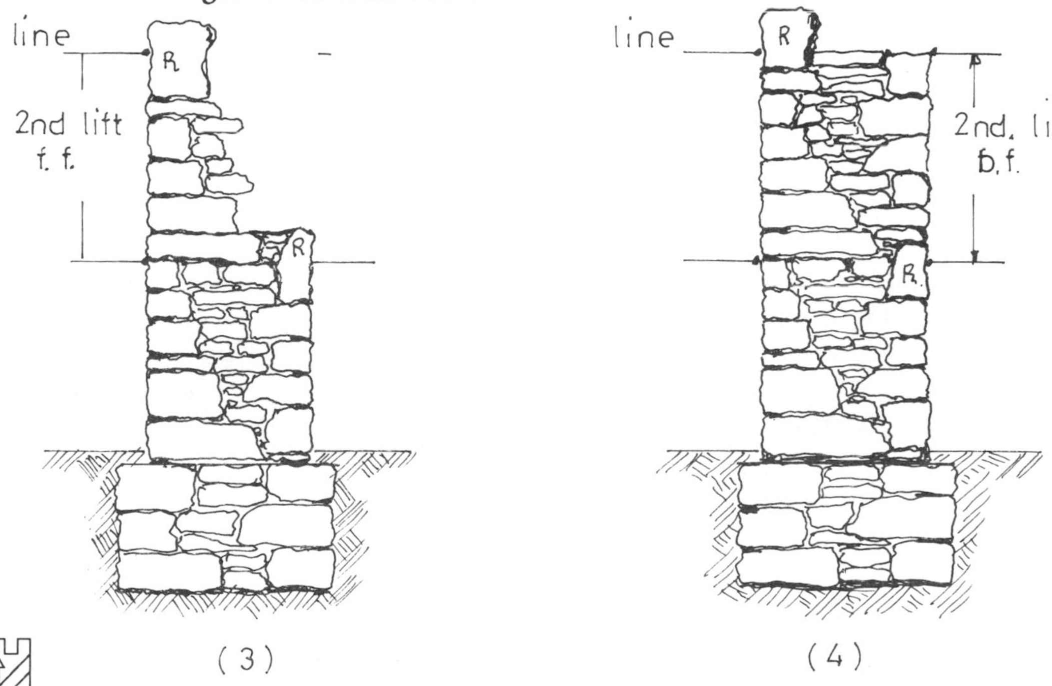

line

R

2nd lift
f. f.

R

(3)

line

R

2nd. lift
b. f.

R

(4)

Figure 9 shows how to tie over or bond the two faces of the wall together by building up the front face and working up and tapering at the back. Then do back face and tie over. Infill with small stones and your mortar mix. Do not use concrete or strong cement.

Figure 10

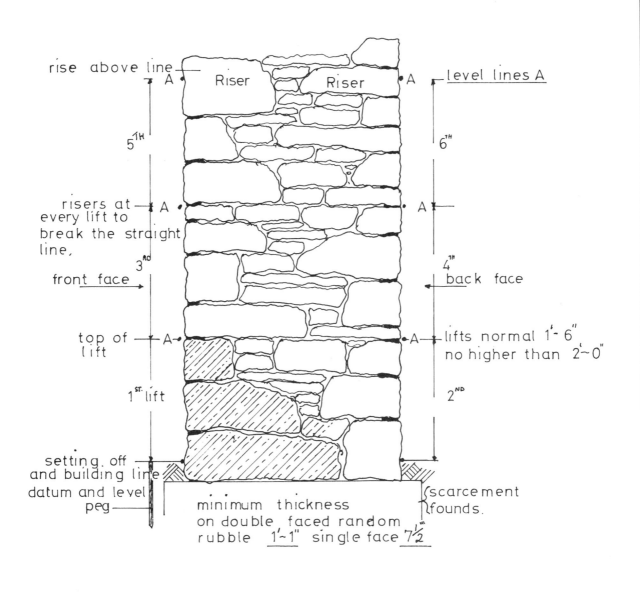

Method used in building double faced random rubble

rise above line — A • Riser • Riser • A — level lines A

5TH 6TH

risers at — A •   • A
every lift to
break the straight
line.

3RD 4TH

front face →   ← back face

top of — A •   • A — lifts normal 1'- 6"
lift   no higher than 2'- 0"

$1^{ST.}$ lift 2ND

setting off
and building line
datum and level } scarcement
peg — } founds.

minimum thickness
on double faced random
rubble 1'-1" single face $7\frac{1}{2}$"

Figure 11

DOS AND DONTS IN RUBBLE BUILDING

allow min. 1-6" or 460 mm
before riser

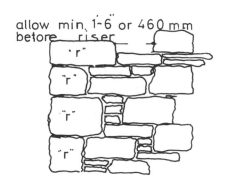

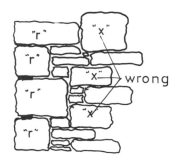

right method

never start rubble rising
above level of rybit. "r"

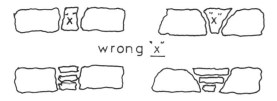

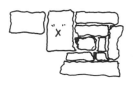

wrong "x"

right method.

never put in a stone
higher than its length.

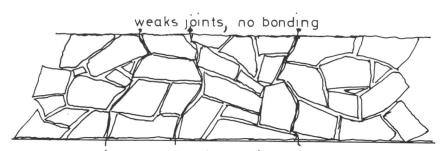

weaks joints, no bonding

this style of building is not masonry, makes a
very weak structure, depends on a strong mortar mix
for bonding

All bed joints must be arranged at right angles to the
pressure coming upon them. This applies to all kinds
of masonry.

Figure 12

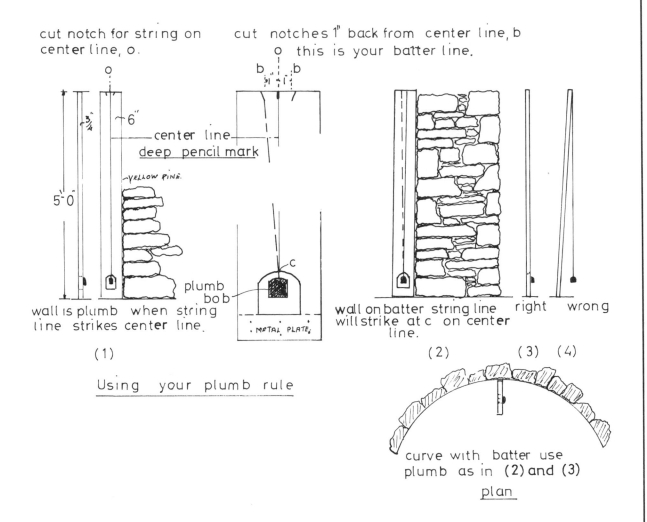

cut notch for string on center line, o.

cut notches 1" back from center line, b this is your batter line.

o — 6"

3/4"

center line deep pencil mark

YELLOW PINE.

5'-0"

wall is plumb when string line strikes center line.

plumb bob

c

METAL PLATE

(1)

Using your plumb rule

wall on batter string line will strike at c on center line.

right wrong

(2) (3) (4)

curve with batter use plumb as in (2) and (3)

plan

This is the old-fashioned style of plumb rule and the most accurate. When you are plumbing the wall, and the string line with the bob on the bottom strikes the center line, it is plumb. To find your plumbing mark every time on the same point, especially on random rubble, mark with chalk, for batter, or slope on wall face, shifting line to notch, b. When string line touches point c, you have a batter.

Figure 14

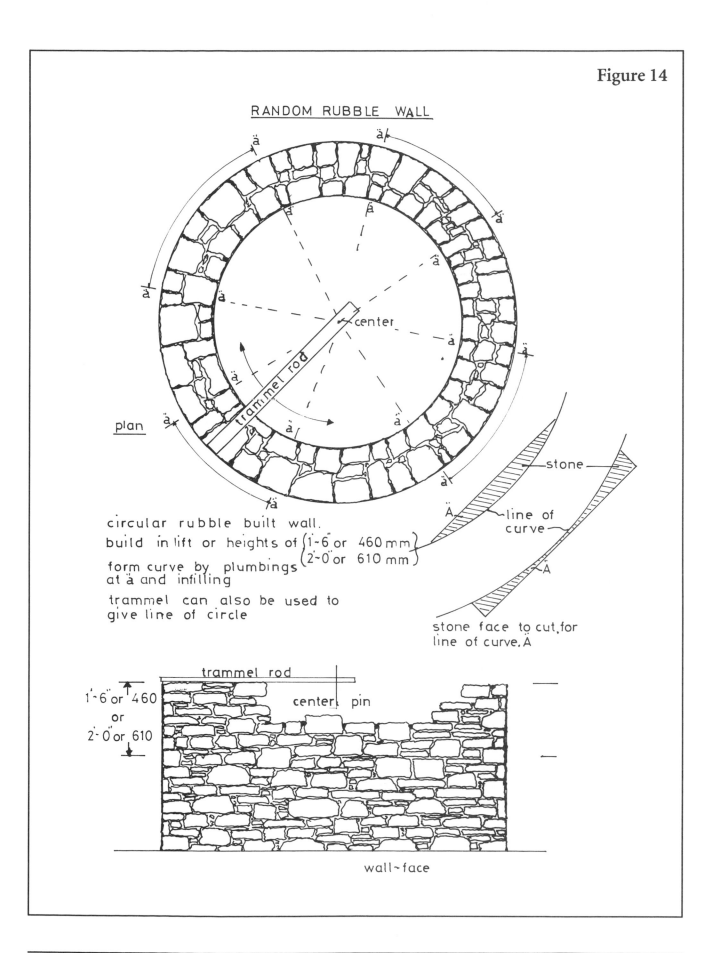

RANDOM RUBBLE WALL

center

trammel rod

plan

circular rubble built wall.
build in lift or heights of (1'-6" or 460 mm)
(2'-0" or 610 mm)
form curve by plumbings
at ä and infilling

trammel can also be used to
give line of circle

stone

line of
curve

stone face to cut for
line of curve, Ä

trammel rod

center pin

1'-6" or 460
or
2'-0" or 610

wall-face

Figure 15

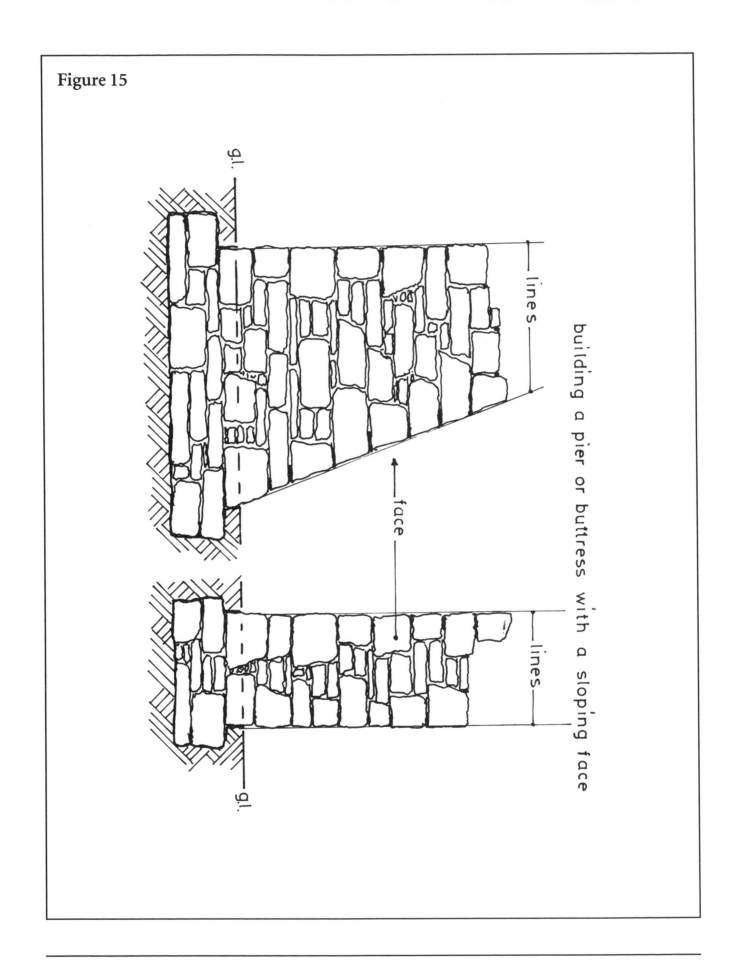

Figure 16

cock and hen ä

but and ben b

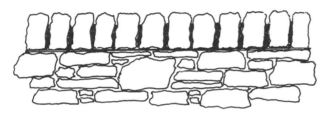

hammer dressed č

do not flush point ä b̈ č

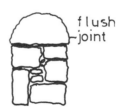

segmental hammer finish

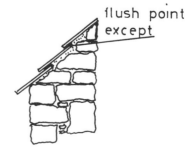

split stone

Figure 17

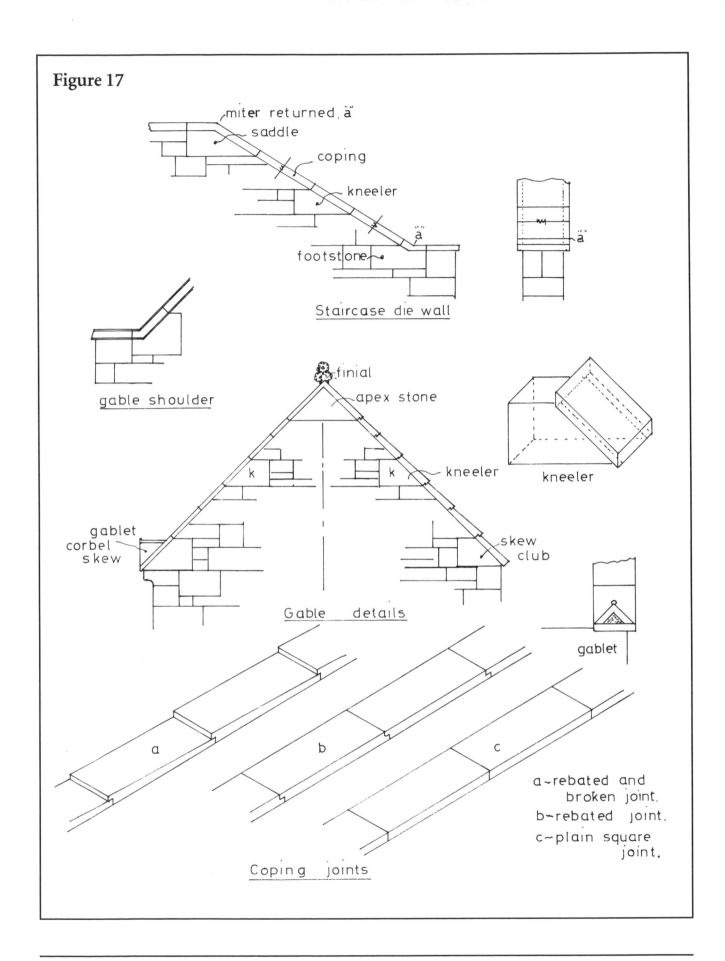

miter returned a''

saddle

coping

kneeler

a''

footstone

a''

Staircase die wall

gable shoulder

finial

apex stone

k   k   kneeler

kneeler

gablet corbel skew

skew club

Gable details

gablet

a   b   c

a~rebated and broken joint.

b~rebated joint.

c~plain square joint.

Coping joints

Figure 18

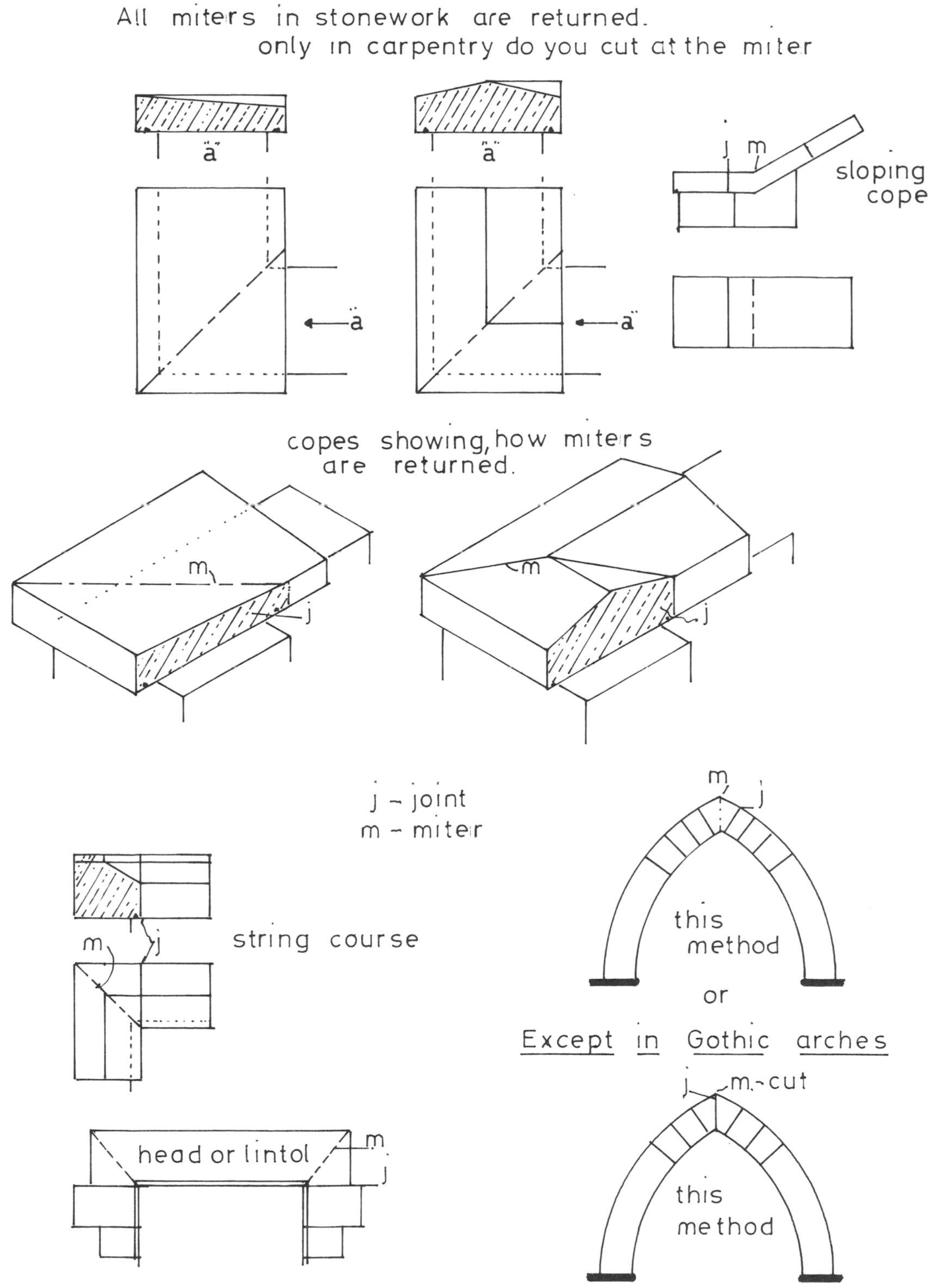

All miters in stonework are returned.
only in carpentry do you cut at the miter

ä"
ä"
sloping cope
← a
← a"

copes showing, how miters are returned.

m
j
m
j

j – joint
m – miter

m
j
this method
or
Except in Gothic arches
j
m – cut
this method

m
j
string course

head or lintol
m
j

Figure 19

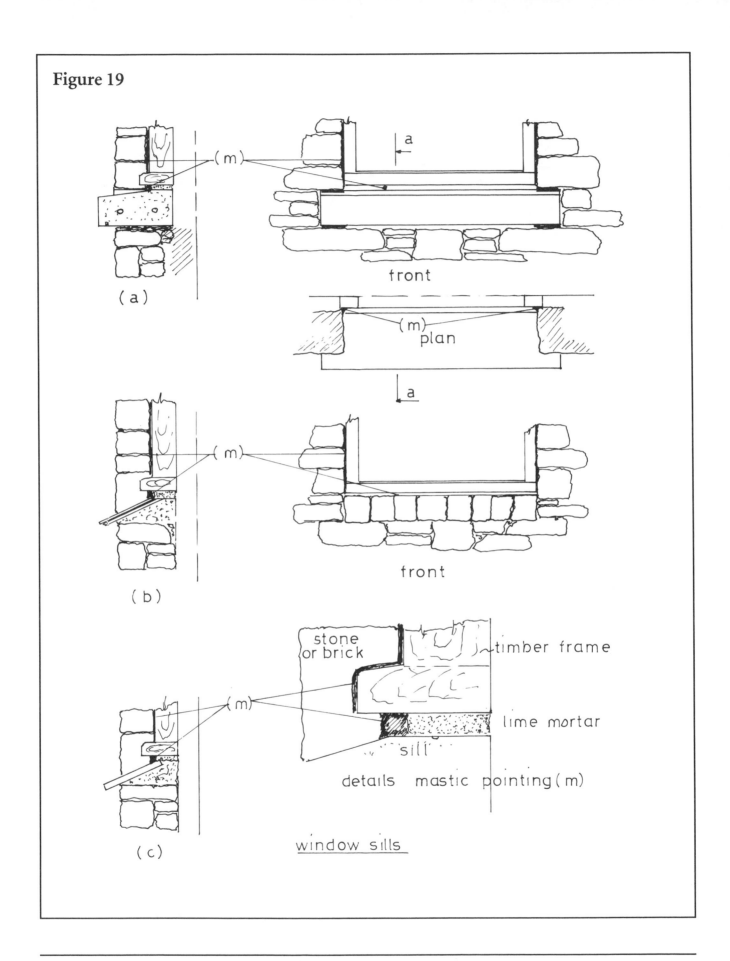

(a)

front

(m)
plan
a

(b)

front

stone
or brick

timber frame

lime mortar

sill

details mastic pointing(m)

(c)

window sills

Figure 20

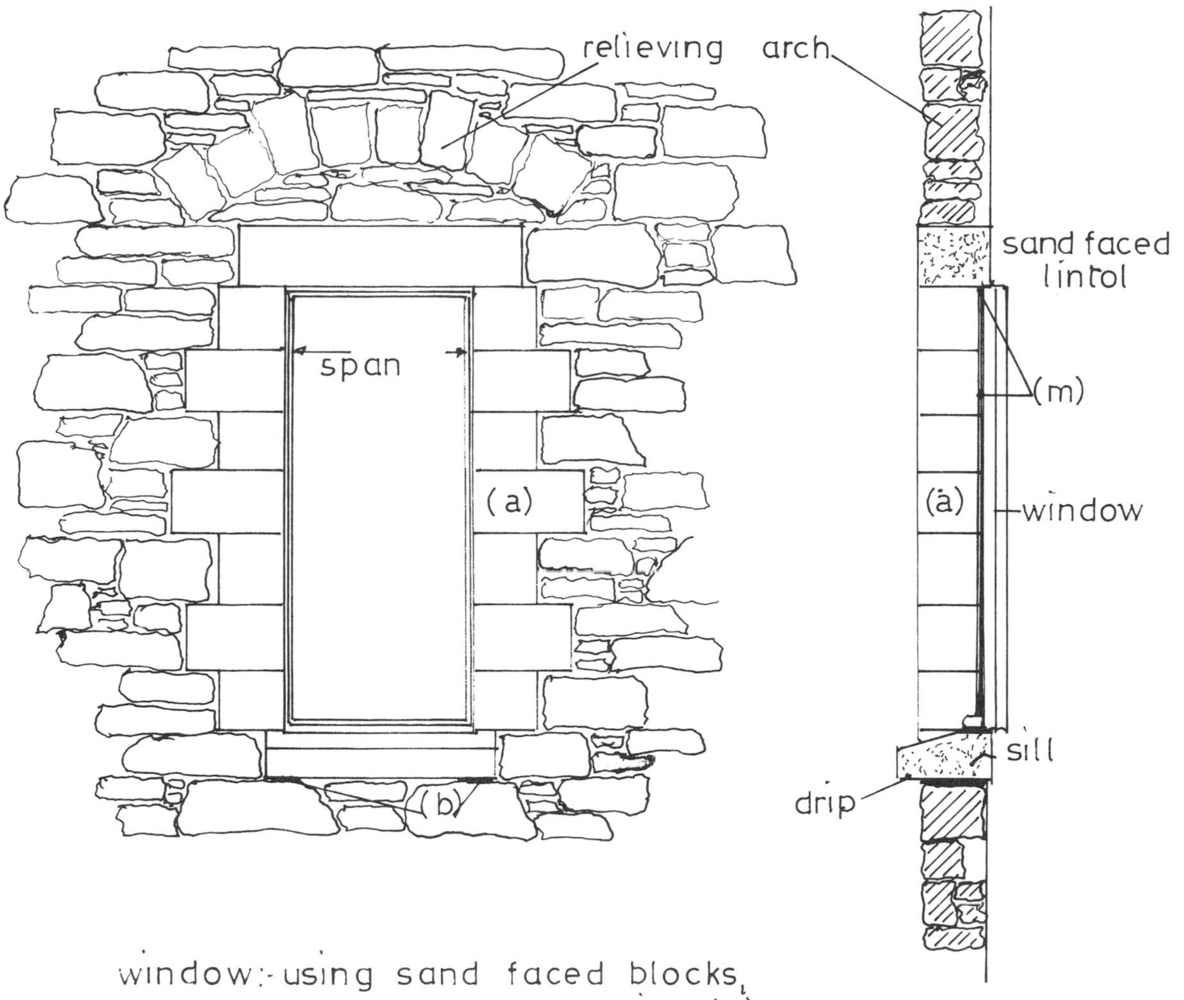

relieving arch

sand faced lintol

(m)

window

sill

drip

span

(a)

(b)

(m) — lime

sill

window:-using sand faced blocks,
instead of dressed stone quoins (a)

relieving arch required if span is
over 3'-0"

oil-sand mastic pointing between
wood and stone (m)

sill-mortar at ends only. (b)

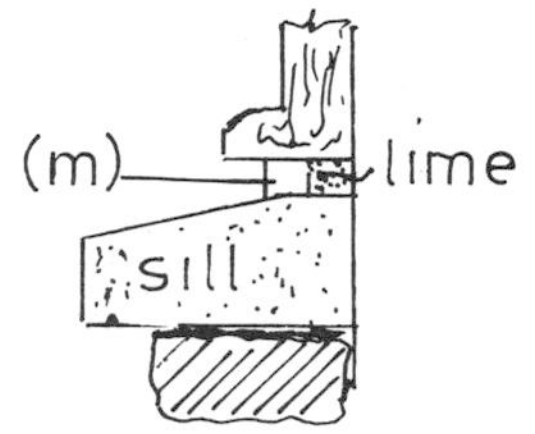

Figure 21

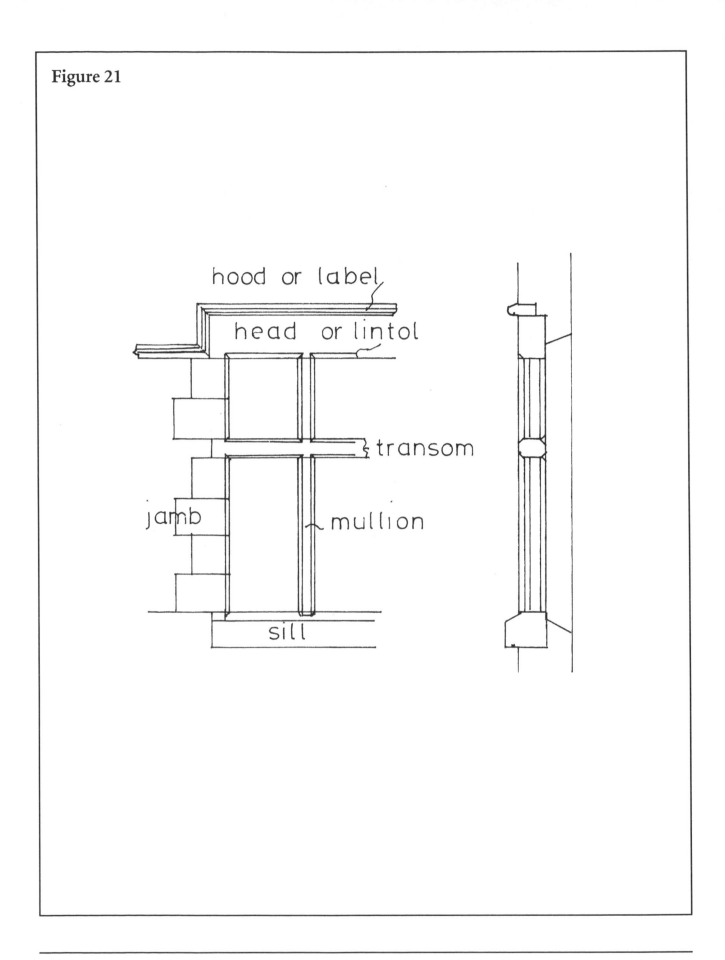

Tower Construction

The Tower's Foundation

The tower is built using the methods described for building a random rubble wall.

For the foundations, remove the total area of topsoil, going down to the hard subsoil. You should average about 1 foot deep, keeping it level throughout and making your excavations about 9 feet in diameter. (The tower in Figure 22 is 8 feet in diameter.) In Figure 22, the base C is 3 feet high from ground level to the first string course and has a 4-inch *batter* (the upward and backward slope of the outside face).

Mark the center point in your excavations and set in the ground a 1-inch diameter metal pipe ("d") to use as a guide for your trammel rod ("T"). It is most important that this be plumb and resist any movement. Use 6-foot lengths of pipe, with the ends adapted to add extensions as needed to increase the height.

With the first pipe set in position d, you will need a trammel rod $1/2$ inch thick, 3 inches wide, and 5 feet long. Mark a point 4'4" from the end of the rod. Drill a 1-inch diameter hole at this point to fit over the pipe, allowing the trammel rod to slide up and down and turn on the pipe.

The end of the trammel rod will act as the plumb line to the bottom of the base. Make a saw cut 4 inches in from the end of the rod. This will give the batter line for the base (see base plan, Figure 22). It will also act as the plumb line "a" on the main structure when you get past the first string course.

Set the trammel 3'1" above ground level and place a clamp on the pipe below the trammel rod. This keeps the trammel from sliding down the pipe. At the other end, the batter line should come through the saw cut up to the top of the pipe. Tie the string ("s") around a peg and jam it into the top of the pipe. To keep the trammel level, tie a knot on the string to prevent it slipping through the saw cut.

Build the stonework solid through the base course, building in the pipe, and building up to the string course. The circle and batter are formed by turning the trammel around and building to the line (see base plan, Figure 22).

Once you have built up to the top of the base course, set your string course (any thickness), cut to suit the circle, and make sure it is level.

Continuing to Build

As you continue, refer to Figure 22, plan at B.

Add an extension pipe to pipe d and raise the trammel rod 3 feet up the pipe, following the process described above. This time the batter line is the plumb line (see plumb line "a"). Build this up as shown in Figure 14 on page 31.

The circular wall will be 12 inches thick, so you should make a gauge to determine the thickness. Cut a strip of wood 12 inches long and hold it against the outside plumb line. Build up the center wall at the same time, as shown on Figure 22, plan at B. Build all the stonework in 2-foot high lifts.

You must also increase the thickness of the external wall where the landing will rest. The center wall functions as one part of this. The center part either can be built solid or you can use half bricks as a backing. It is easier to form the inner circle this way. Section "a" shows the landing resting on a brick backing.

The height of the tower is governed by the number of steps it will have. Allow a rise of $7 1/2$ to 8 inches, with a $1/4$-inch run on each tread. Build in the steps once you have determined the finished height of the steps, plus the last rise onto the landing.

Plan top A shows the steps and landing. You can raise the height of the step to 9 inches to gain more height. This is normal in a turret stair in a castle.

Finishing

To finish the top of the tower, set your string course at least 3 inches below the top of the finished landing, allowing for a $1\frac{1}{2}$-inch overhang. Set the trammel rod above the finished parapet height, and adjust the plumb line out $1\frac{1}{2}$ inches. At this point, the wall will be $13\frac{1}{2}$ inches thick, so you batter this up to 12 inches from the inside of the parapet wall. The castellated parapet height is 2'6" from the top of the string course to the lowest point on the lower castellation, with a back slope of 9 inches; the high castellation is 9 inches above low point "L" and 9 inches above high point "H".

Each high and low section of the castellated parapet is 3 feet wide on the outside face (see plan top A). The side face angles are lines radiating from the center point. The covering on the parapet, in this case $\frac{1}{2}$-inch thick Blue-stone, is cut to fit the curve, bedded in 3 parts sharp sand, 1 part lime, and 1 part cement. It is finished with a coating of boiled linseed oil to bring out the color and provide extra waterproofing.

Steps are cut from $1\frac{1}{4}$-inch thick stone slabs, 13 inches wide at the wall side and 2 inches wide at the center point. They are all "winders," with a rise of $7\frac{1}{2}$ inches.

Build the steps in solid as you go up, making sure that your steps "rise," and finishing one step below the landing.

Set the landing on the prepared bearing walls, bedding solid with a flush point. Finish by cleaning and painting with boiled linseed oil.

For the doorway, see the discussion of flat arch construction on page 68.

I created this tower design from memory, from some of the old buildings I have worked on in Scotland. The drawing was made after the work was completed.

The Tower, built August 1990, in Easton, PA. It was designed from memory of restoration work in Scotland and built by the author and his sons. Random rubble with limestone string course and stone winder stair up to half landing. The entrance has a flat arch.

Tower doorway completed, with hood above arch and continuation above string courses.

Tower up to setting level for string course.

Random rubble stone boundary wall, showing sloping copes and small piers with stone bells.

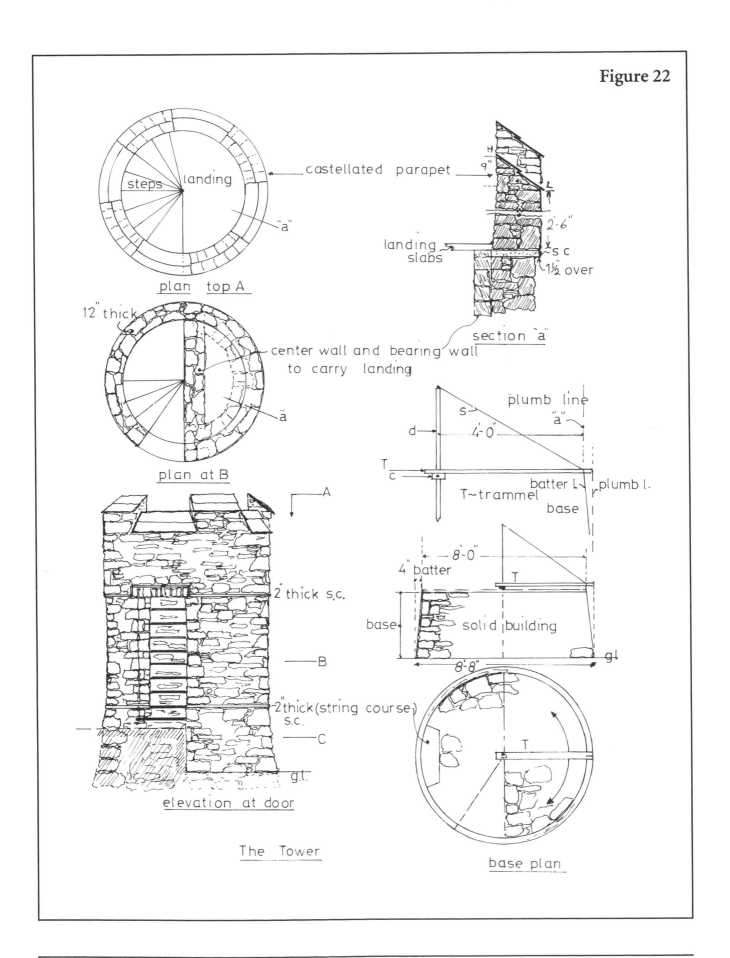

Figure 22

plan top A

castellated parapet

9"

2'-6"

landing
slabs

s c

1½ over

section "a"

12" thick

center wall and bearing wall
to carry landing

"a"

plan at B

plumb line

s'

d

4'-0"

"a"

T
c

batter l.

plumb l.

T~trammel

base

8'-0"

4" batter

T

base

solid building

8'-8"

g l

2" thick s.c.

A

B

2" thick (string course) s.c.

C

g.l.

elevation at door

The Tower

T

base plan

Fireplace Construction

Building a fireplace in rubble or round river stone can cause difficulty in setting up, in keeping your stone in position while plumbing the ends, and in arching over. I will explain a simple method I have used for this type of fireplace.

Before you build, find out if the position you have chosen was the site of an existing fireplace, whether there is a hearth, and whether there is a chimney vent to build into.

All sizes given are normal for coal- or wood-burning fireplaces in the United Kingdom. For use in the U.S., most readers will prefer a wider fireplace opening. The sizes can be increased to suit your own requirements.

If your fireplace is sited where none was before, you will need to build, or have built, the chimney and hearth first (Figures 23, 24, 25, and 26).

We will make our fireplace 6'6" long by 3'6" high, from floor level. The fireplace opening is 18 inches wide by 22 inches high (from the top of the proposed hearth). Make sure you mark the finished hearth level. For our example, the hearth is 4 inches thick.

From your lumber yard, you will need two pieces of 1" x 6" pine 3'6" long for the end plumbings ("a" in Figure 27). You will also need a piece 1" x 6" and 7'0" long for the top piece to hold your two plumbing boards in position ("b"). Nail or screw, making sure all is level, square, and plumb, adding side braces to hold it in position. This frame forms the outline of your fireplace and acts as the plumbing for the ends and face.

Make a frame for the fireplace opening, 18 inches wide inside and 26 inches high (22 inches plus the hearth thickness of 4 inches). Make sure the frame is square and level. Fit the frame in position, making sure it *lines through* (the frame is not twisted) with the end boards that are acting as your plumbings. When setting the frame for the opening, set it on small wedges. When the fireplace is complete, it is only a small matter of removing the wedges and your frame should drop down, making it easier to remove.

Now that your frames are in position, set your building line "L" (as shown) 12 inches up from floor level. Stretch this line from end board to end board, making sure the line is level at all times. Start by building up to your line, as described in the section on random rubble work, keeping your snouts to the line. Once you reach the line, on both sides take a small piece of wood and tamp all your joints in gently. This tightens the mortar and gives depth for the finishing point. It also keeps the face of the stone clean. You should repeat this tamping process every time you reach the line.

Your next lift on the line should be level with the top of the board spanning the opening. Now build up to this level, allowing the end stones to project above the line ("c"), as in a riser. Having reached this height, and with the tamping process complete, leave the work for about twelve hours before carrying on. This allows the sides to dry out before you apply the top load.

Raise the line 9 inches for the next level. At this level we form the arch over the opening. First, spread damp sand over the board spanning the opening. This acts as a cushion to set the stone on. Tamp the sand until it is firm, allowing a rise at the center of the board of 1/4 inch. Now build up at "d". This is where you set your first arch stones. Use a thin stone for the arch, as large stones on a small fireplace spoil the effect of the arching. Set the first arch stones at either side, then fill in between the arch stones and your plumbing ("e"). Once this is completed, proceed to form the arch, working in from both sides until it meets in the center. For your arch stones, you can add an extra trowel of cement to your mortar mix and make it damper for easier working. The joints should be tighter on your arch stones.

Complete the infill, building between the top of your arch stones and the underside of your top board. The back of the stone face can be infilled upon removal of the board.

Your stonework on the fireplace is now complete. Leave it for twelve hours before you start removing the outside frame boards. Then gently clean and tamp your joints. Leave the opening framed in until the framing around has been removed. Then carefully remove the wedges and the frame should drop down, clear of the underside of the arch. Tamp and clean the joints. The fireplace is now complete, other than the final point. Avoid using a strong mortar mix. For the final pointing mix, I recommend 6 parts sand, 1 part lime, and 1 part cement. For your building mix, use 4 parts sharp sand, 1 part lime, and $1/2$ part cement.

In forming your hearth, building a stone face, and infilling, set your finishing slabs on top. Do not use concrete; use your building mix.

Points to Observe

- Your frames, when set in position, must be level, square, and plumb.

- Overlap your stones, avoiding long vertical joints, as this weakens the structure. Do not overload by trying to build too high in one day.

- Tamp the joints lightly after each lift to the line.

- Do not use any metal bars or lintols in the construction of a fireplace. All metal expands with heat and will cause cracking.

- Add wire ties only if required, in the back wall of the fireplace.

- A flat arch span of a maximum 2'6" should be used in this type of fireplace. A larger span requires a *segmental arch*.

- Do not use strong mortar mixes or concrete.

Figure 23

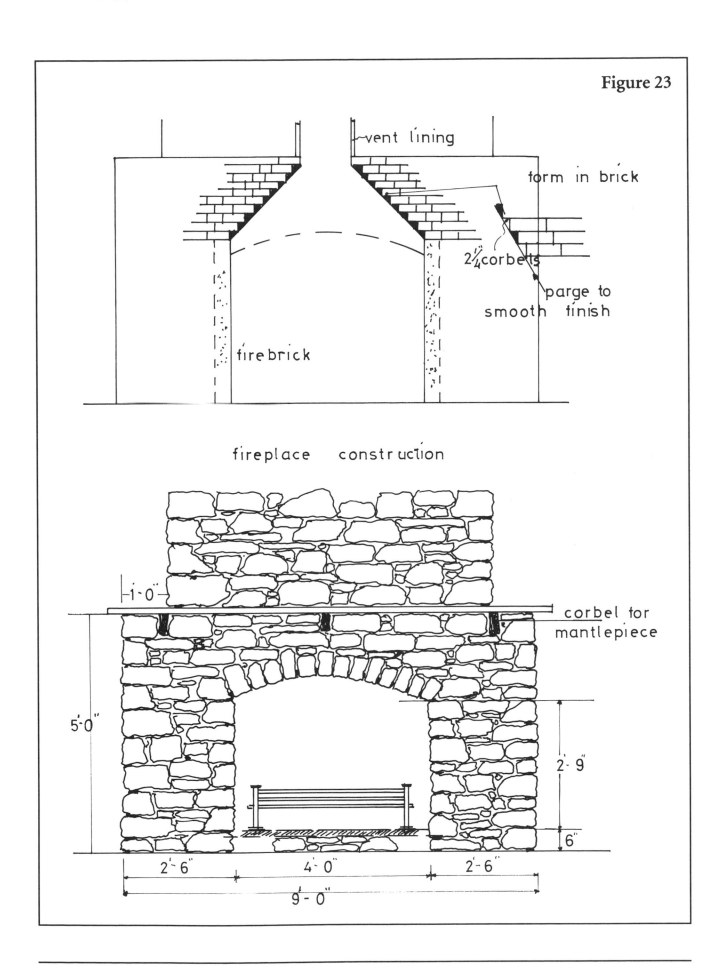

vent lining

form in brick

2¼" corbels

parge to smooth finish

firebrick

fireplace construction

1'-0"

corbel for mantlepiece

5'-0"

2'-9"

6"

2'-6"   4'-0"   2'-6"

9'-0"

Figure 24

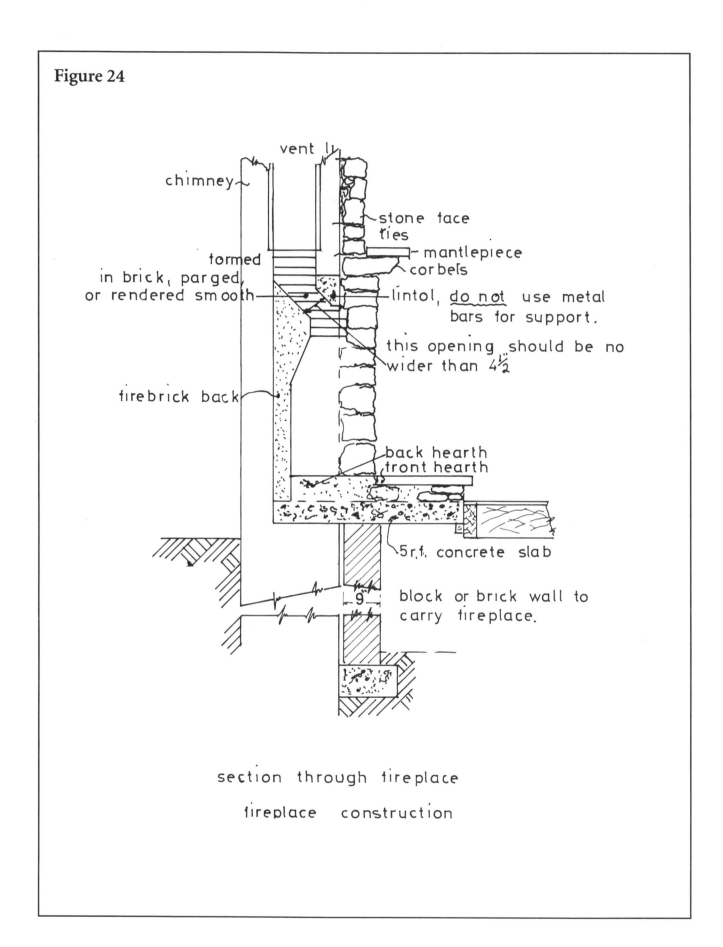

vent

chimney

stone face ties

mantlepiece corbels

formed in brick, parged, or rendered smooth

lintol, <u>do not</u> use metal bars for support.

this opening should be no wider than $4\frac{1}{2}$

firebrick back

back hearth
front hearth

5 r.f. concrete slab

9"

block or brick wall to carry fireplace.

section through fireplace

fireplace construction

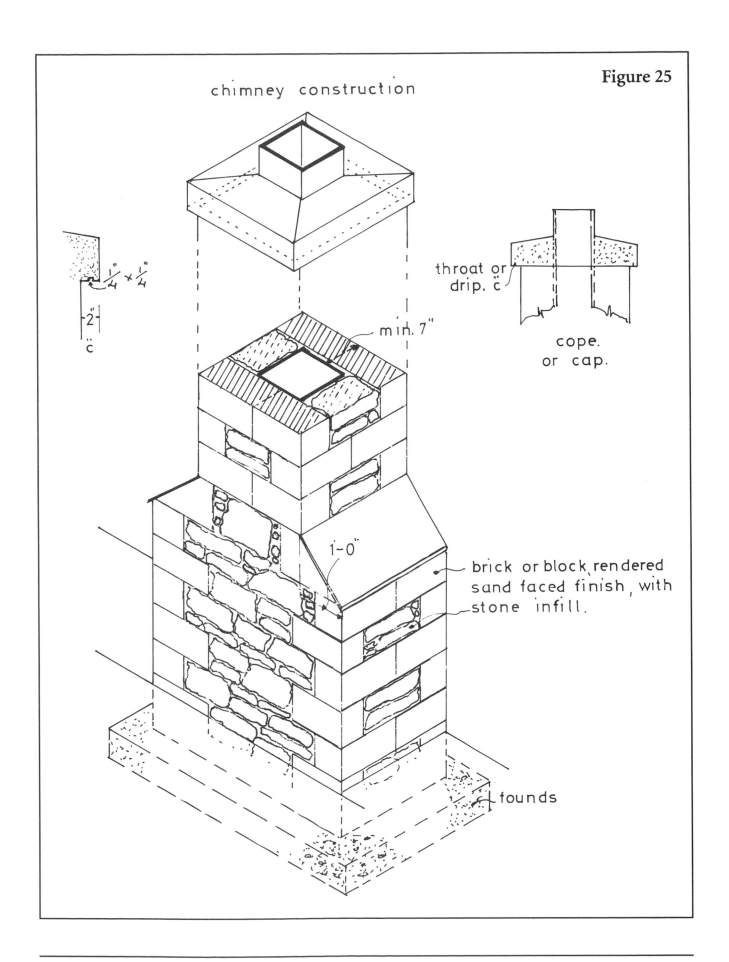

chimney construction

$\frac{1}{4}" \times \frac{1}{4}"$

2" c

min. 7"

throat or drip. c

cope. or cap.

1-0"

brick or block, rendered sand faced finish, with stone infill.

founds

Figure 25

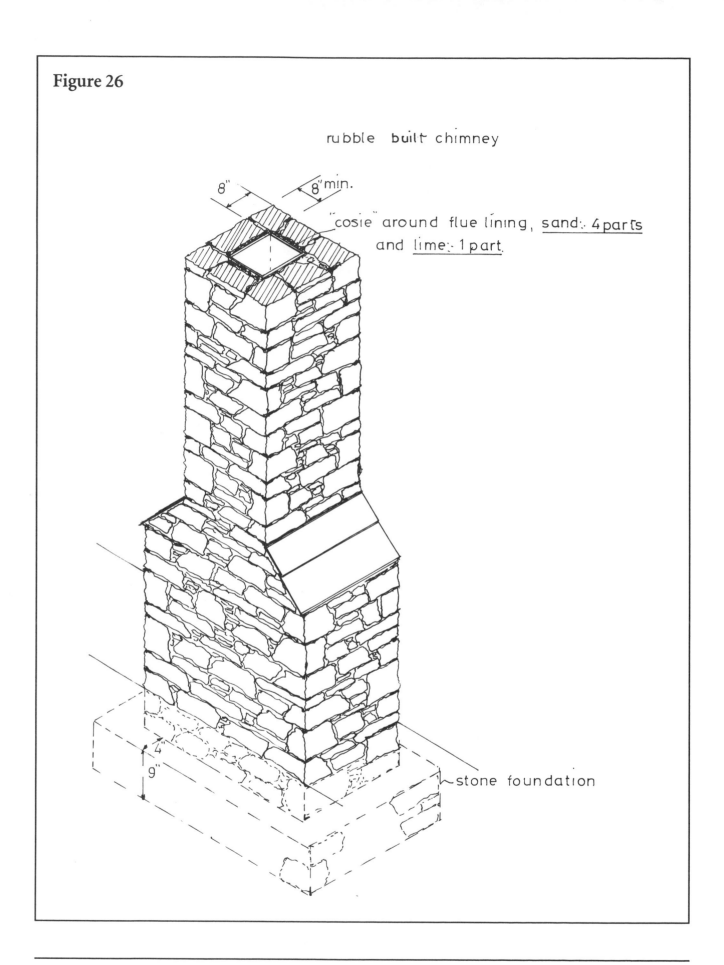

Figure 26

rubble built chimney

8"

8" min.

"cosie" around flue lining, <u>sand:</u> 4 parts
and <u>lime:</u> 1 part.

stone foundation

4

9"

Figure 27

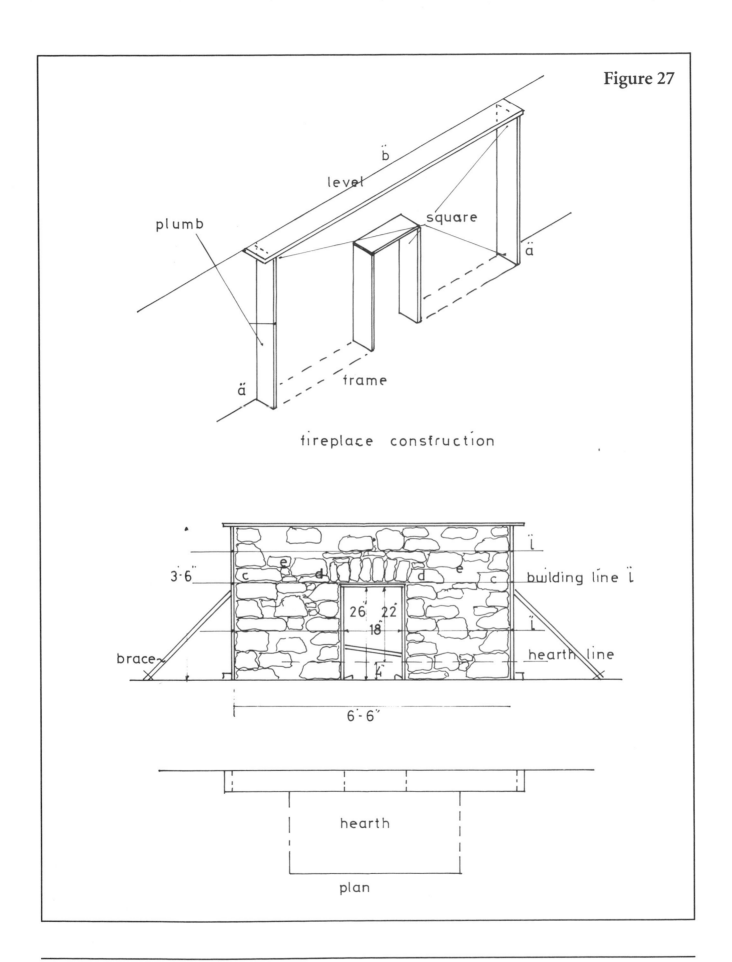

fireplace construction

Stairs

Stone Stairs

Stairs are composed of blocks of dressed stone, so fixed as to be a means of ascending from one story of a building to another, or from one level of a garden to another. Here are some definitions you may find useful:

Staircase — The room or space containing the stairs.

Tread — The upper surface of the step; that which gives foothold.

Rise — The vertical height between two treads.

Nosing — The edge of the tread, sometimes moulded and projecting.

Soffit — The underside of stairs.

Flight — Steps comprising a continuous series without a landing.

Fliers — Steps comprising a flight, rectangular in plan.

Winders — Steps tapering in plan. Used in place of a landing.

Landing — There are two kinds of landing: half space and quarter space (Figure 28).

Other general information you should know:

- You should have no more than twelve risers or steps before a landing.

- The tread of a step should not be less than 9 inches or more than 12 inches, and the rise no more than 7 inches or less than 5 $\frac{1}{2}$ inches.

- A rule generally adopted for tread and rise is to multiply them together to achieve a product of 66; i.e., 11-inch tread x 6-inch rise = 66. The

---

Technical Details — Stairs

For the following discussion, refer to Figure 29.

There are three kinds of stairs: those fixed at both ends; those fixed at one end with the other end free, known as hanging steps; and steps on a circular plan either with a central newel, which forms part of the steps, or with an open well. Steps secured at both ends are fixed as the wells are built. Hanging steps may be fixed the same way, or spaces may be left in the wall to receive them. The wall in which the hanging steps are built should be strong mortar 18 inches above and below the steps. The ends are pinned into the wall with slate or tile, and run with cement grout, care being taken that the joints are tight and even. The free

ends are supported until the cement has set. When supported at both ends, the steps have a 6-inch wall hold. Hanging steps must be built 9 inches into the wall. Sections of square steps are shown in Figure 29. The method in which the steps simply rest on each other is the most common form. Any of the other sections are better used where each step is rebated or checked onto the one below, to prevent sliding.

Hanging steps generally have the best appearance and are usually *spandril*-shaped, with the underside splayed or worked away, and finished either plain or moulded. Hanging steps get a good support from the wall but depend chiefly for stability on the support given by one step to the other.

result may be one more or less than 66, but the rule is a good one to use as a guide.

- In setting out a stairway, the length to be taken for the flight and the height to be gained should be marked out on story rods (lengths of 1" x 3"). The number of steps and their proportions are obtained from the above rules. The rods on which the heights are marked are used in fixing the steps.

Circular Stairs

Circular stairs are built in a circular chamber. When they have a central newel, one end of each step is fixed in the wall, and the other end is worked in a circular plan to a 3-inch radius. This forms a continuous pillar up the center of the staircase. Circular stairs with an open well are built on the same principle as hanging stairs, but all the stairs are winders. To gain headroom, the steps are pitched high. This means an increase in the step riser (the vertical height between two treads), allowing for a steeper slope on the soffit (underside) as a spandril step. This increases the angle of pitch, giving more headroom.

Building External Staircases

The general method for building external staircases is that used for random rubble building. (Refer to Figures 30, 31, and 32.)

Use precast concrete treads or existing stone cut treads (Figure 30). The side walls will act as rests for your steps. As you build them up, set each step in position. The steps should not rest on

Random rubble staircase with die wall.

each other. Leave a $\frac{1}{8}$-inch joint, and overlap $\frac{1}{4}$ inch. Once this joint is completed, point with a fine mix of 3 parts sand, 1 part lime, and 1 part cement. Having built up the side walls, infill with rubble and tamp down to allow for drainage. Do not use concrete for your mortar mix.

Build two side walls and a front wall (Figure 31, "a"). On the front wall, leave an open joint in the stonework for drainage. After you have built the three walls as shown in the drawing, infill with gravel or small stones up to the level of the walls, making sure it is tamped in solidly. Cover it over with a good, thick bed of 6-1-1 sharp sand mix. Set your 1 $\frac{1}{2}$-inch or 2-inch thick slab on this mix and tamp down to let it settle into the mix. Allow $\frac{1}{4}$-inch run on the tread for every foot of run.

With the first step now set, proceed with the next step using the same method, only carry your front wall over your slab about 1 inch. This saves water penetration on what would be a very awkward joint to point. Also, allow your slabs to overlap $\frac{3}{4}$ inch to allow for drip. Point the slab joints with a mix of 3 parts sand, 1 part lime, and 1 part cement.

Infill that has been tamped down allows any moisture to drain away, whereas a solid, hard infill such as concrete allows moisture to lie on it, causing the slab to crack or lift during frost weather.

Your *wall rests* (the top part of the wall on which the steps rest) should be double faced and a minimum of 1 foot thick. If they have more than five steps, build the inner face with brick or block, allowing 7 inches minimum for the stone face.

Random rubble staircase with die wall.

Figure 28

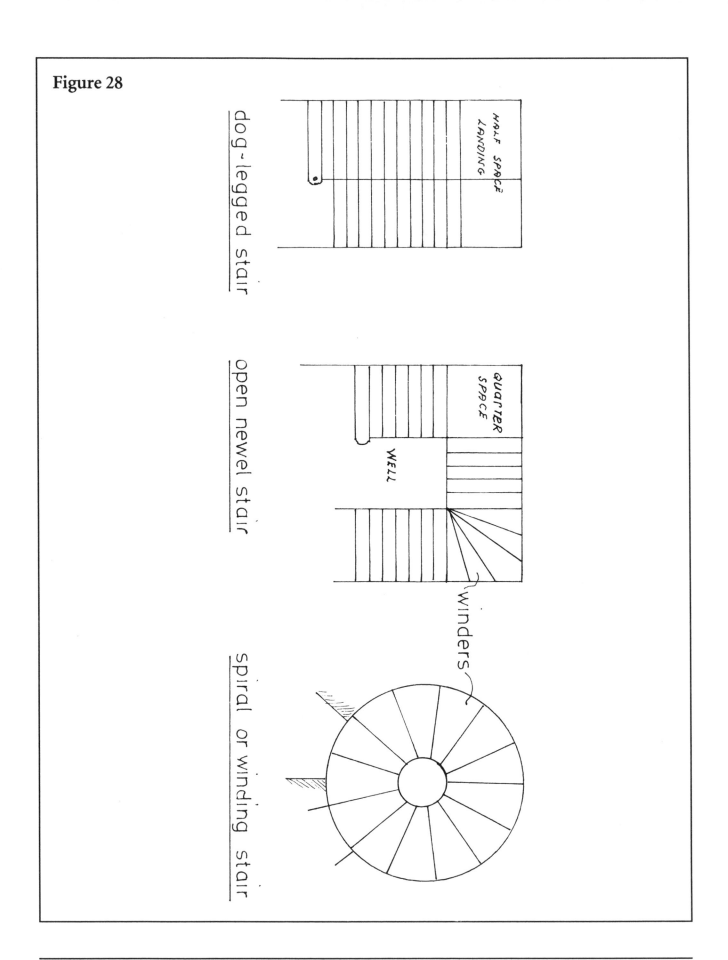

dog-legged stair

HALF SPACE LANDING

open newel stair

QUARTER SPACE

WELL

winders

spiral or winding stair

Figure 29

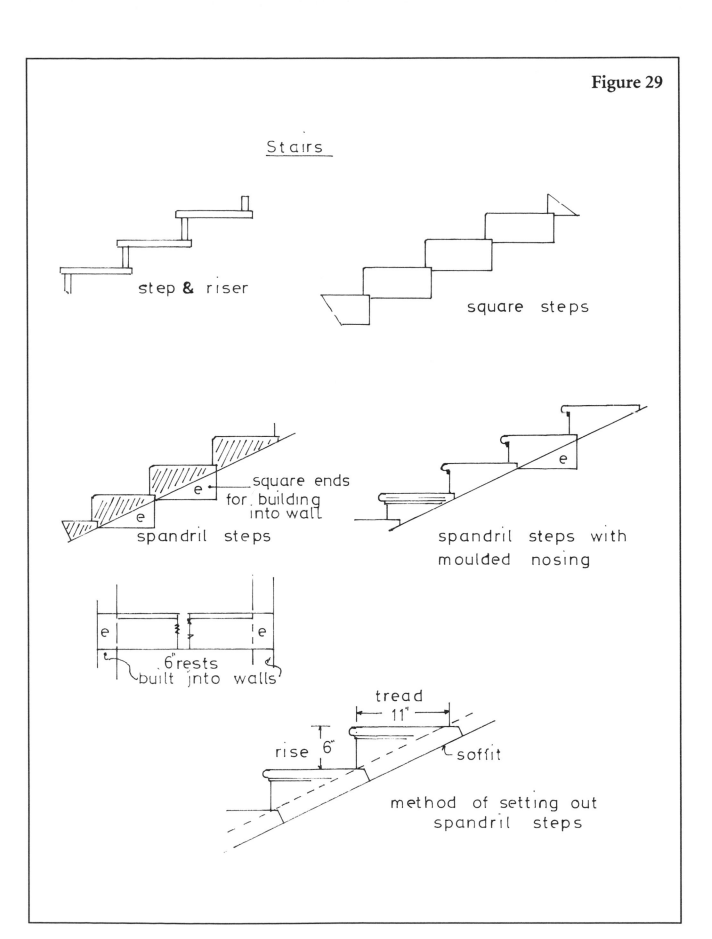

Stairs

step & riser

square steps

spandril steps

square ends for building into wall

spandril steps with moulded nosing

6" rests built into walls

tread 11"

rise 6"

soffit

method of setting out spandril steps

Figure 30

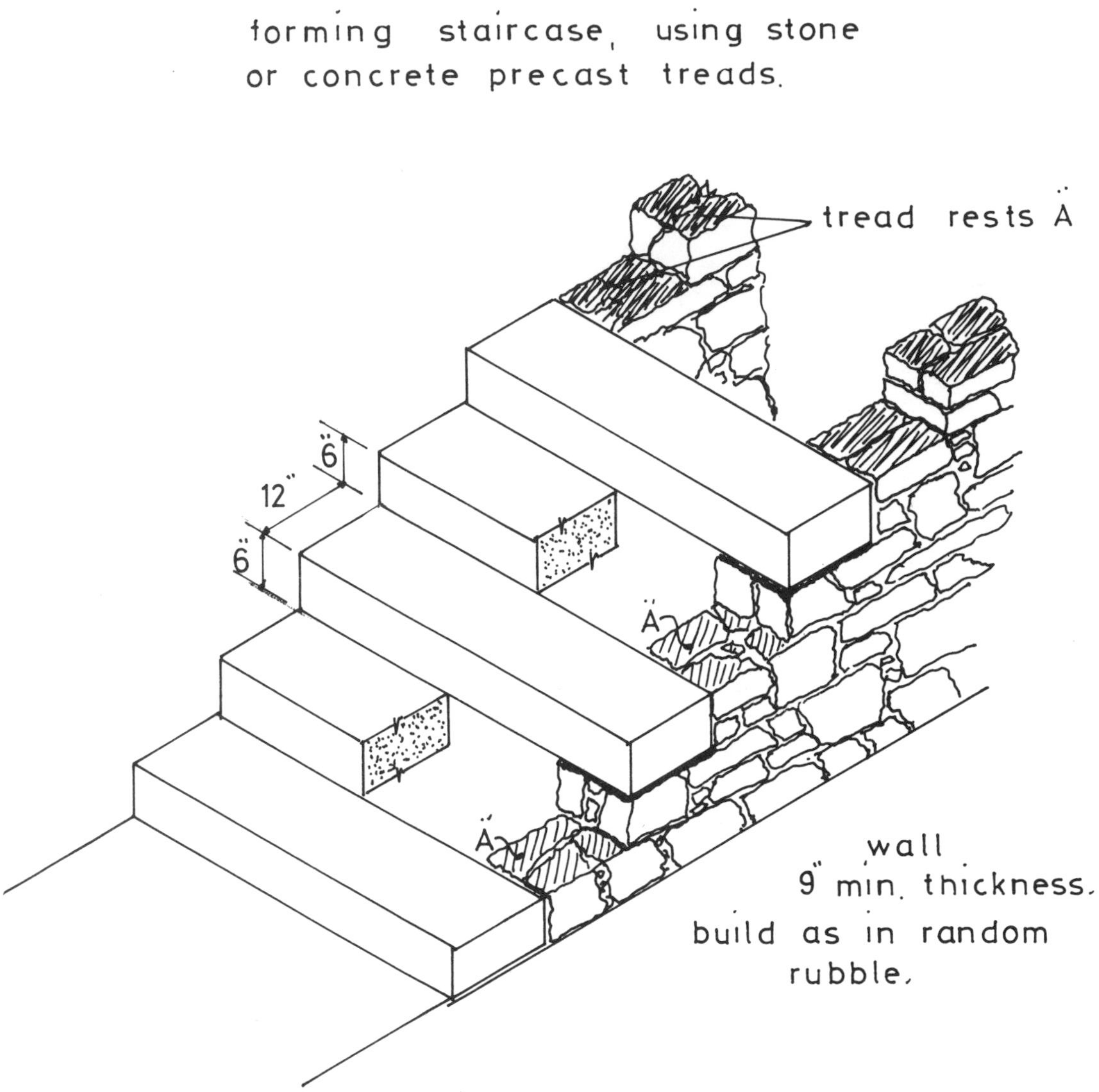

forming staircase, using stone
or concrete precast treads.

tread rests Ä

6"

12"

6"

Ä

Ä

wall
9" min. thickness.
build as in random
rubble.

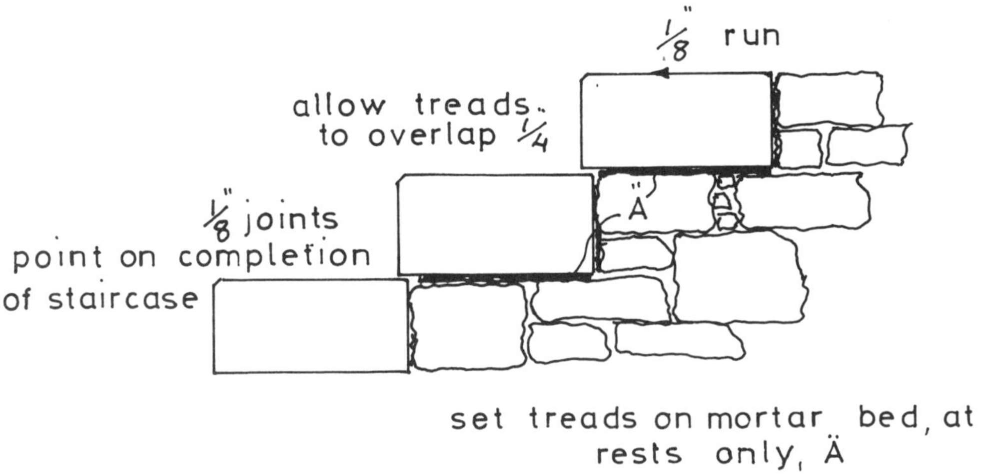

1/8" run

allow treads.
to overlap 1/4"

Ä

1/8" joints
point on completion
of staircase

set treads on mortar bed, at
rests only, Ä

Figure 31

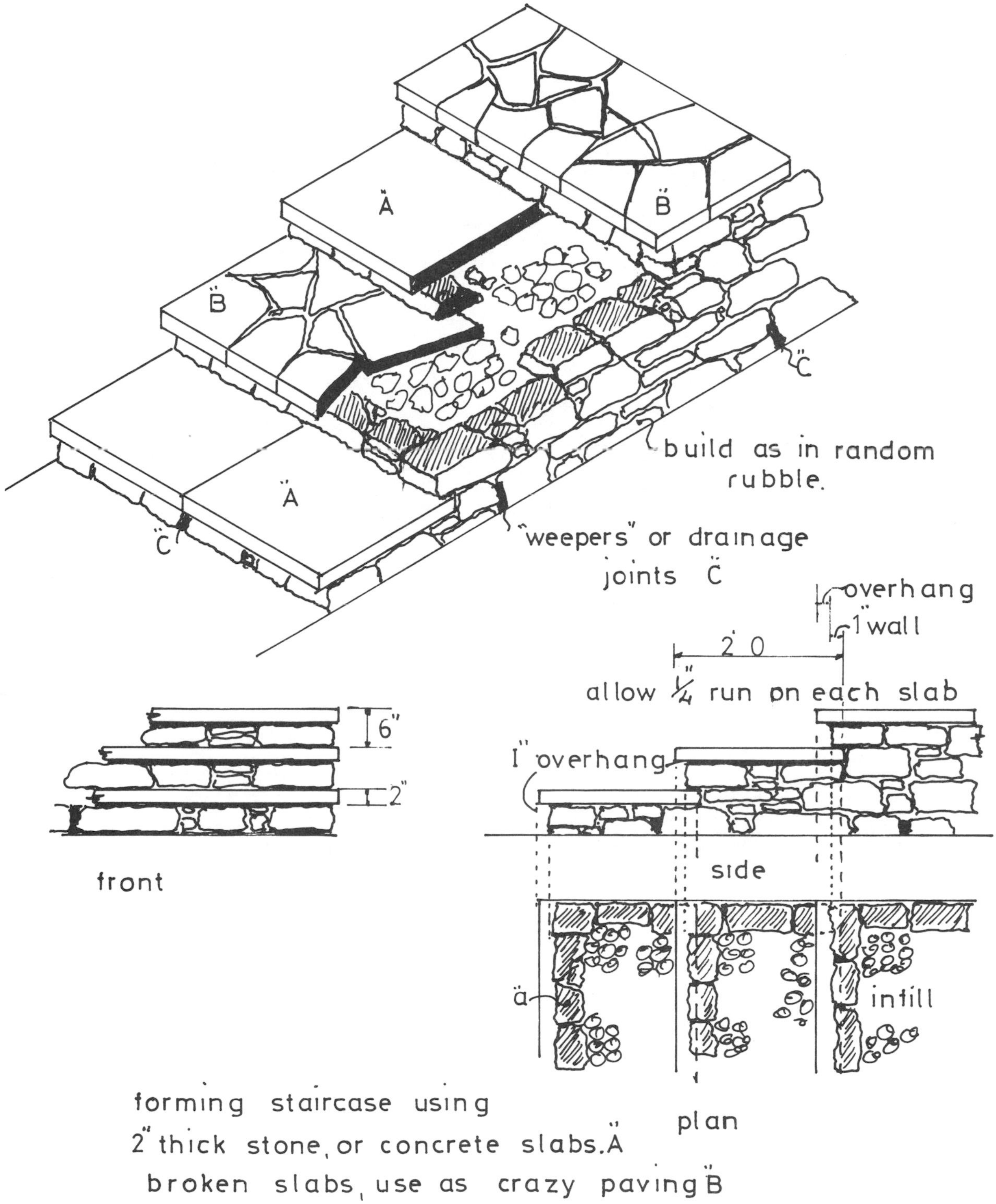

build as in random rubble.

"weepers" or drainage joints C̈

overhang

1"wall

2 0

allow ¼" run on each slab

I" overhang

6"

2"

front

side

ä

infill

plan

forming staircase using
2" thick stone, or concrete slabs. Ä
broken slabs, use as crazy paving B̈

Figure 32

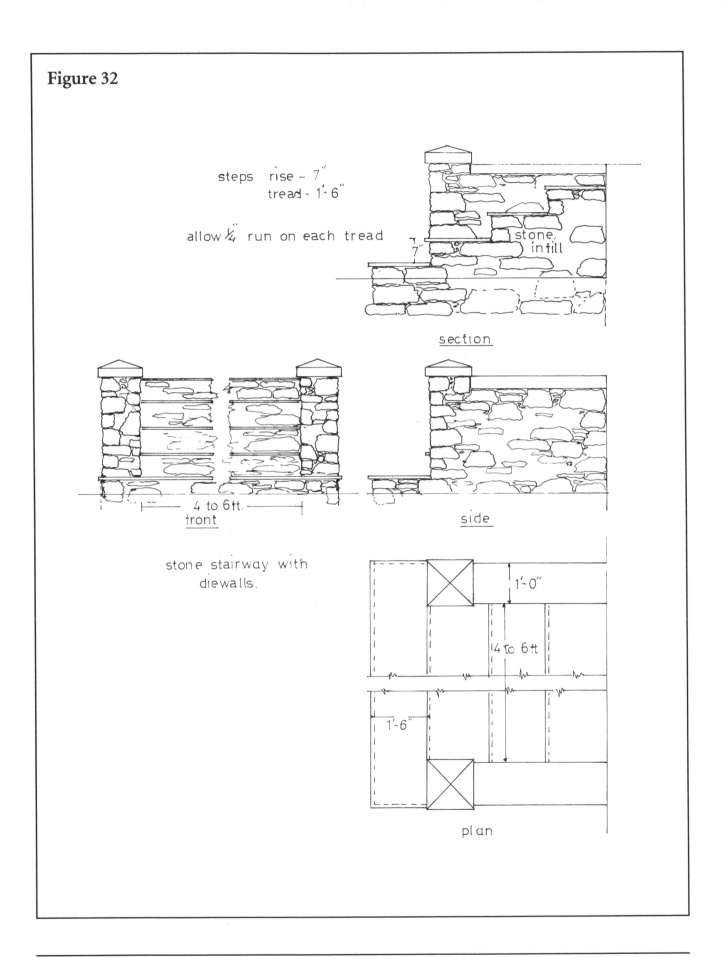

steps rise – 7"
      tread – 1'-6"

allow ¼" run on each tread

7"

stone infill

section

4 to 6 ft.
front

stone stairway with diewalls.

side

1'-0"

4 to 6 ft

1'-6"

plan

Arches and Their Construction

Arch Construction

Arches in architecture are composed of a number of stones arranged symmetrically over an opening intended for a door, window, etc., for the purpose of supporting a superincumbent weight. Arches are named for the shape of the curve on the underside, or *intrados*, and are either simple or complex. Simple arches are those that are struck from one center, or any segment of a circle. Complex arches are those struck from two or more fixed centers; among these are many of the Gothic or pointed arches.

Types of Arches

Semi-circle or Saxon arch — (Illustration 1, Figure 33.) "A" and "B" are the span and "C" the center. From point "C", the curves are struck and the joints radiate. The center "C" is on the springing line of the arch. The beds are horizontal.

Segmental arch — (Illustration 2, Figure 33.) This arch has "A" and "B" as the span and "C" the center for describing the curve, with the end of the trammel in "C". The distance "CA" describes the line of the *intrados*. Set off the depth of the *voussoirs* (arch stones) and, with the trammel set at "C", describe the line of the *extrados*. From "A" and "B", draw lines radiating to "C", which gives the line to the joints of the arch. Divide the intrados and extrados into as many parts as you want to have stones in the arch. Radiate all the lines to "C", which will give the proper direction of the joints. In a segmental arch, the radius is regulated by the height and span.

Segmental arch with three centers — (Illustration 3, Figure 33.) Divide the springing line "AB" into four equal parts. The first points "C"

and "D" of division from each springing point are two of the centers. For the third center "E", make each end division "AC" and "DB" the base of an equilateral triangle, with one extended side of each meeting on the center line at "E", giving the center of the circle.

Semi-Gothic arch — (Illustration 4, Figure 33.) The intrados is a semi-circle. The extrados is struck from two centers "AB" on the springing line, depending on the face of the arch at the springing line and at the crown. The centers "AB" are found by drawing lines from points "C" and "D" and bisecting these lines. The intersections with the springing lines are the centers.

Modified Gothic arch — (Illustration 5, Figure 33.) The rise of this arch is less than the span. To find centers, draw a line from the springing point to the top of the rise. Bisect this line, and at its intersection at "C" on the springing line is one of the centers. The other center "D" is found in the same manner. All arches in which the rise is less than the span are known as "depressed" or "drop" arches.

Lancet or pointed Gothic arch — (Illustration 6, Figure 33.) The rise is greater than the span of this arch. The centers are found as described for the modified Gothic arch.

Four-centered pointed arch — (Illustration 7, Figure 33.) Divide the springing line "AB" into four equal parts. The first divisions, "C" and "D", from each springing point create two of the centers. The other two, "E" and "F", are on the perpendicular lines at a distance from the springing points equal to the span. The rise in this case is two-fifths the span.

Equilateral arch — (Illustration 8, Figure 33.) The centers "A" and "B" are struck from each springing point, with the span as radius.

Points to Observe in Arch Construction

- There must be sufficient weight and strength in the abutments to resist safely the overturning moment of arch thrust.

- There must be sufficient area in the piers and arch to prevent failure by crushing.

- The thrust of an arch, a tendency all arches have, is the quality of descending in the middle and overturning or thrusting asunder the points of support. Note: The amount of thrust of an arch depends on the proportions between the rise and the span. The span and weight to be supported must be a defined quantity. The thrust is diminished in proportion as the rise of the arch is increased. The thrust is increased in proportion as the crown of the arch is lowered.

- The joints of an arch are the lines formed by the adjoining faces of the voussoirs. These should radiate to some definite point, and each should be perpendicular to a tangent of the curve at each joint.

- In all curves composed of arcs of a circle, a tangent to the curve at any point would be perpendicular to a radius drawn from the center of the circle through that point. Consequently, the joints of all such arches should radiate to the center of the circle of which the curve forms a part.

Making Arch Centers

The erection of a suitable centering for the support of arches during their construction is of great importance for their ultimate success.

The worker, whether a carpenter or a stonemason, should be acquainted with the methods of fixing centering and should be able to state what is required for any given piece of work.

The centering should be sufficiently strong and rigid to resist effectively any deforming stresses. The form and construction are varied to suit the requirements of each individual case.

Owing to the increasing load, the voussoirs, or arch stones, are gradually built up from the springers, or lowest point of the arch. There is a tendency for the pressure to raise the crown and depress the haunches of the centering. This must be braced to resist this pressure effectively.

The center should also be rigid when fixed in position, with provision made for striking. This minimizes the risk of damage to the stonework when the centering is removed. The center should rest upon folding wedges (also called fox wedges) on top of upright supports. When the arch is complete, the folding or fox wedges may be removed, thus allowing the center to drop clear of the arch soffit (horizontal underside).

Centering for Gothic or pointed arches must follow the correct curve line of the arch. It must be accurate in size and curve line so that the voussoirs may be laid directly on the surface of the lagging pieces, which should be placed only a few inches apart.

Segmental or semi-circular arches should have the lagging pieces nailed on the centering at intervals to suit the sizes of the voussoirs, preferably two lagging pieces for each voussoir. The curve line of the centering, including the lagging pieces, should be made 1 inch less than the arch line radius, thus allowing room for the folding wedges to be used between the lagging pieces and the voussoir soffit.

A radius rod or trammel should be fixed to the centering for testing the accuracy of the curve line of your arch. It should also be used for setting each voussoir so that the joint radiates accurately to the center point. Test each voussoir for plumb, using your plumb line, as shown in Figure 34.

The successful fixing of all centering, and the dismantling for reconstruction of all arch work, requires great care and accuracy on the part of the craftsman involved, who must have a thorough knowledge and skill in this class of work.

In the reconstruction of an arch, before dismantling an arch takes place, you must accurately measure as much of the arch as you can, especially the span, the rise, and the width of angle joints. A center frame is required for demolition, using the same center frame for rebuilding.

Pitched faced squared rubble, with dressed arched window, built 1830.

Arched windows, buttresses, and squared cut ashlar, built around 1200.

Semi-circular random rubble arch with a span of 6'0". Built by Ewan Cramb.

For semi-circular or segmental center frames, allow 1 inch clear of the arch soffit for the folding wedges. For Gothic and other arches, the center acts as your guide. The soffit of the voussoirs rests on lagging pieces, so this must be accurate.

Technical Terms Used in Arch Work

For the following discussion, see Figure 35.

Abacus — The moulded slab crowning the capital of a column, by which the arch or lintol is supported.

Abutments — Those masses of stonework from which the arch springs and to which the superincumbent weight over the arch is transmitted. Exception: When two arches spring from the same mass of stonework, the latter is a *pier*.

Arcade — A series of arches usually in the same plan, supported on columns.

Capital — The moulded or carved cornice of a column.

Center — The point or points from which the curve or curves framing the intrados is struck.

Column — A pillar of circular section.

Cornice — A projecting moulding.

Crown — The highest point of extrados of an arch.

Entasis — The convex swelling on a column.

Extrados or back — The upper or convex side of an arch.

Flank wall — A side wall.

Haunches — The sides of the arch on the extrados from the lowest point on the extrados to about halfway up towards the crown.

Impost — The upper part of a pier or an abutment, on which the arch rests and from which it springs, generally finished with a moulded cap.

Intrados or soffit — The under or concave side of an arch.

Jambs — The sides of abutments or piers.

Key — The uppermost or central stone of an arch.

Piers — The intermediate supports of an arcade.

Radiating joints or normals — Lines taken to the centers, from which the circles forming the curve of the arch are struck.

Respond — The semi-column or corbel forming the support of the end arch of an arcade, usually projecting from a flank wall.

Rise — The vertical distance between the highest point of the intrados or soffit and the level of the imposts or springing points.

Skewback — The part of a pier or an abutment that immediately supports a segmental or flat arch.

Span — The horizontal distance between the springing points.

Spandril — The surfaces of the main wall contained by a horizontal line from the crown, a vertical line from the lowest point on the extrados, and that portion of the extrados between these two points.

Springers — The lowest voussoirs on each side of an arch, or the points on each side from which the curve forming the intrados springs.

Springing line — The line where the intrados of an arch meets the abutments or piers.

Voussoirs — Stones composing the courses of an arch.

Cutting Curves and Joints on Voussoirs

For the following discussion, see Figure 36. You should have in hand a small scaled drawing, showing the span and type of the proposed arch. For this example, we will use a semi-circular arch, which is easy to plot out full-size on a flat surface.

1. Mark out your span as indicated on your drawing. Put two pins at "A" and "B" and stretch a line or string or draw a pencil mark between these pins. This is your span, and it is also the springing line of your arch.

2. Make a dead center line or point between "A" and "B". Put in pin "C" and use this center pin to form your curve by attaching a piece of string as your trammel. Hold a pencil or other marker at point "A" on the trammel. Pull this

A front entrance showing a completed segmental arch with random rubble building. Note the staircase and the stone pier at the door entrance.

The rear entrance with segmental arch, with flat and relieving arches on side window. The arch has random rubble flank walls.

around as shown on the drawing to point "B". This will give you the required curve.

3. Place a piece of thin plywood or sheet zinc, about 3' x 3', on the curve line and, using your trammel, mark the curve line on the plywood or zinc. Keep the plywood or zinc in position (nail it to the floor if this is practical), as your completed arch will depend on accurately drawing out your curve line and joint lines.

4. Extend the trammel, marking a straight line on the plywood, radiating from center point "C". Now you have the true angle of your voussoir joints.

5. Remove the plywood and cut exactly to the marks you have made. This is the face mould for your voussoirs.

6. Use the same method in working out your curve line for the center frame.

Constructing Rubble Arches

The Flat Arch

For the following discussion, refer to Figures 37 and 38.

The flat arch has a small rise to the center of the opening. If you kept it level, it would look on completion as if it were dropping in the center, hence the reason for the rise. Also, if any settlement of the arch did occur, this would tighten it.

Any opening spanning more than 3 feet, using a flat rubble arch or stone lintol, requires a relieving arch. This arch takes the pressure or load off the flat arch or lintol.

For a flat arch, you must build the rubble work up to the points shown at "a" and set up 2" x 6" timbers, leaving about 1/4-inch clearance at each end ("b"). Your timber frame is now set in position. Assuming you have selected the stone to form your arch, mix up sand and lime in equal portions — enough to cover the timber ("c"). Make the mixture damp and spread it over the board, allowing it to rise 1/2 inch on the center line of the opening ("d"). This rise is for any span from 2 to 4 feet. For larger spans, allow 1/4-inch rise on every foot run of stone to the center line.

Using a 7-1-1 building mix, add to each bucket of mix one trowel of cement and one of lime. This makes your mixture more "buttery" and gives better adhesion to the stone. Start at ends "e", working in from each end, spreading the mortar

This page and facing page: The front entrance with segmental rubble arch, showing the center frame used to support the arch, as well as the springer, voussoir stones, and spandril infill.

mix on the joints only ("f"). Set your stone on the sand-lime mix, pressing into the mix to steady the stone and hold it in position. Now complete this process working in from both sides, until you are about 2 to 3 inches from the center line ("g").

To fit your center or key stone, spread mortar on each joint surface of the stone already in position ("h"). Lower the stone into the opening — do not use a hammer — until it rests on your sand-lime mix. If joints are too slack, adjust each joint a little to make them all look equal, then flush point the face joints, pressing the mortar in gently. The stones of the arch are now in position, with the face joints pointed. Add a touch of water to your mix, making it into a grout or slurry. Pour this into the back of your arch stones and into any voids in the joints, making sure it does not push out your stones. The lime mix the stones are resting on will prevent the grout from running through. Once your joints are filled up, insert small slivers of stone into each joint, pushing them down gently into the grout, until they are tight. These small stones act as a wedge in each joint.

Leave the work for a few hours then, using a small piece of wood, scrape back ¾ inch of the face joints to allow for the finishing point. Allow at least thirty-six hours before you remove the timber frame. Use a hard brush to remove the sand-lime mix from the underside of the arch stones. Clean down and your flat arch is complete.

Note: DO NOT USE any metal bars, enclosed or exposed, or any strong cement mortars or concrete within or near any stone arches.

The Relieving Arch

The relieving arch requires a center frame constructed as shown in "a", Figure 37. Build it up at ends "b", making these points level across. This forms the seat for the center frame, which you set on your fox wedges as shown ("c"). This allows for easy removal of the center frame on completion of the arch.

Before you set up the frame, nail a thin piece of wood "d" on the center line of your frame, with a small nail projecting. This is the center point of your curve, to form the arch.

Set your frame in position, and tie a piece of string over the projecting nail "e" to act as your trammel or joint guiding line. Once you set the stones forming the arch, hold the string against

A relieving arch.

each of the joints to see if they radiate to the center point.

Set the stones hard on your center frame. Your first stones act as springers — the load carriers. Make sure you build up at each end after you have set the springers ("f"). Use the same mortar mix as described for the flat arch, spreading your mortar across the joint. Then set your arch stones, working up from each end until you reach the center line. Do not build up one side to your center line. Build your stonework up equally from each side. The procedure is much the same as in building a flat arch, except that you do not need to set the stones on a sand-lime mix.

Build up each side until you come to the center, then insert your key stone. Flush point the face joints, gently pressing the mortar well in. Use a slurry mix to fill any voids, with slivers of stone as wedges, pressed into the joints.

Start building your stone at each end ("g"), filling up the spandril of the arch. Do not start at the top of the arch and work down. Once you have completed this area, any load is on the springers. Leave this structure for 24 hours before you attempt to remove the center frame. To remove the frame, tap the wedges until they slacken, then remove the wedges, allowing the center frame to drop $1/2$ inch clear of the bottom surface of the arch. Then the frame can be easily removed. Clean out the space left (the *tympanum* "h") between the top surface of the flat arch and the soffit or bottom surface of the relieving arch, building up until this space is closed. This is the only load being applied to your arch.

A relieving arch with tympanum infill and flat arch.

Building a segmental relieving arch: the frame is set level and wedged into position.

Starting from the springer over the arch frame.

Figure 33

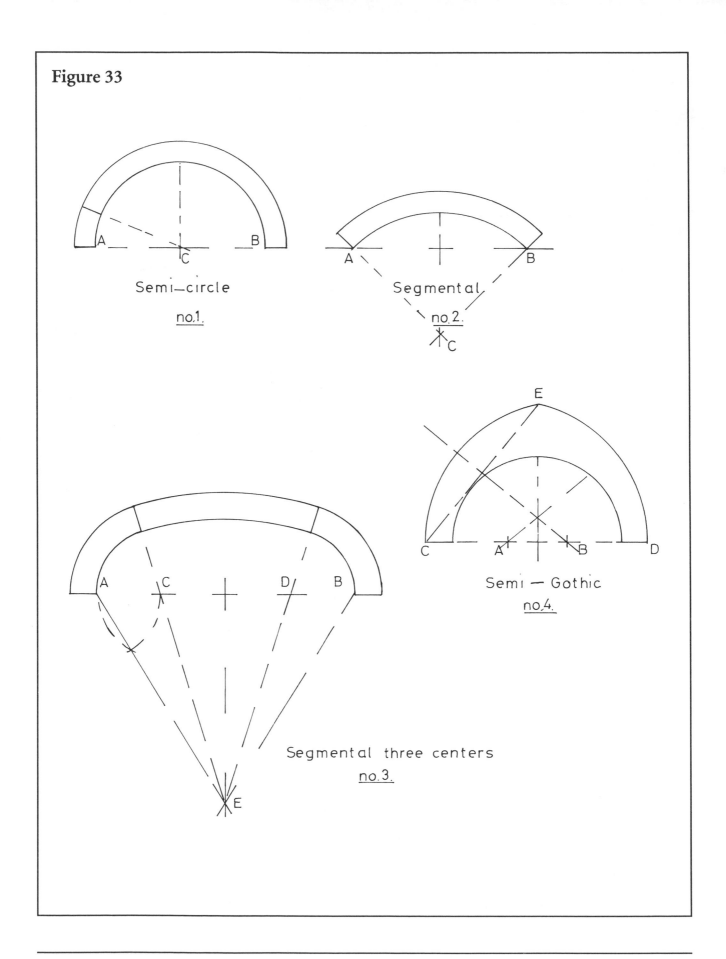

Semi—circle

no.1.

Segmental

no.2.

Semi — Gothic

no.4.

Segmental three centers

no.3.

Figure 33 (Cont.)

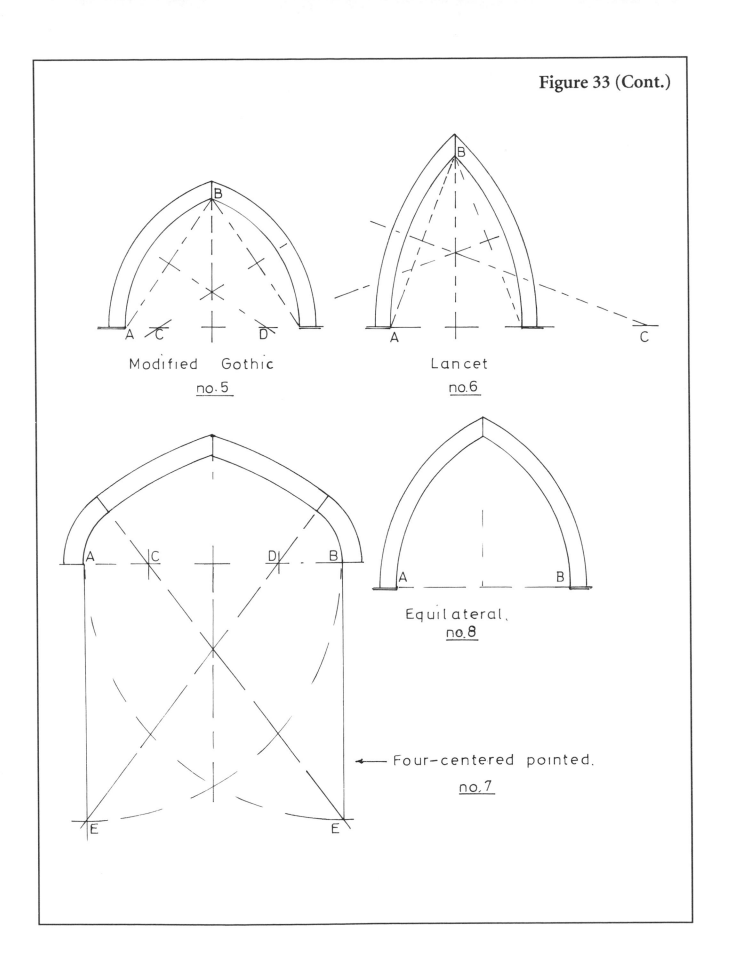

Modified Gothic
no. 5

Lancet
no. 6

Equilateral,
no. 8

Four-centered pointed.
no. 7

Figure 34

ARCH CONSTRUCTION

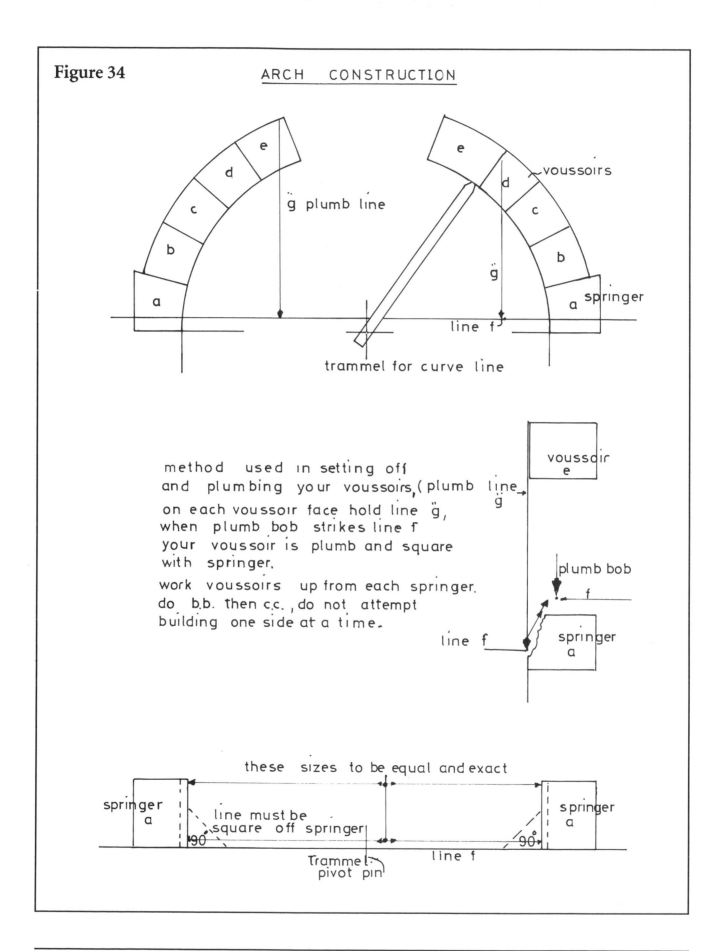

voussoirs

ġ plumb line

ġ

line f

a springer

trammel for curve line

method used in setting off
and plumbing your voussoirs, (plumb line
on each voussoir face hold line ġ,
when plumb bob strikes line f
your voussoir is plumb and square
with springer.

work voussoirs up from each springer.
do b.b. then c.c., do not attempt
building one side at a time.

voussoir
e

plumb line
g

plumb bob

f

line f

springer
a

these sizes to be equal and exact

springer
a

line must be
square off springer

90

Trammel
pivot pin

line f

90

springer
a

Figure 35

ARCH CONSTRUCTION
terms used

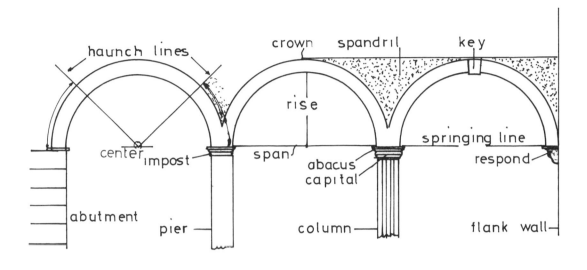

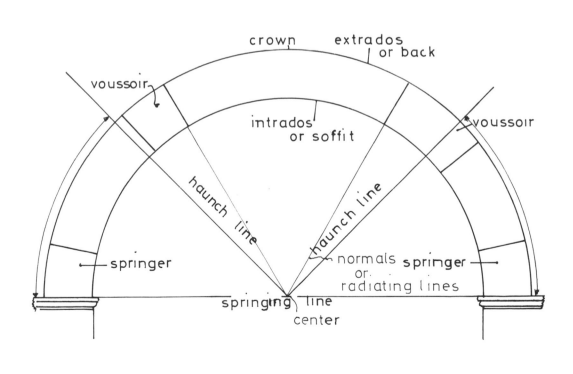

Figure 36

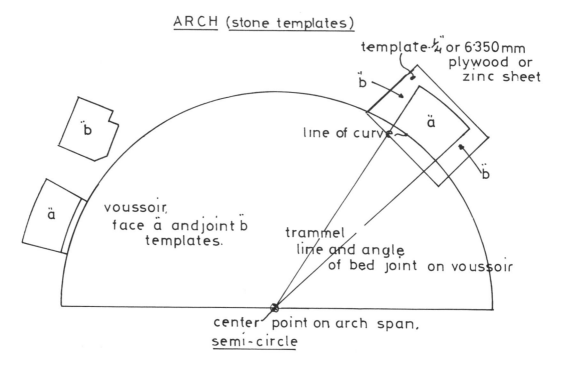

ARCH (stone templates)

template·¼" or 6·350 mm plywood or zinc sheet

b̈

ä

line of curve

b̈

voussoir, face ä and joint b̈ templates.

trammel line and angle of bed joint on voussoir

ä

center point on arch span, semi-circle

method for setting off for a full size working template, for joint and voussoir curve line.

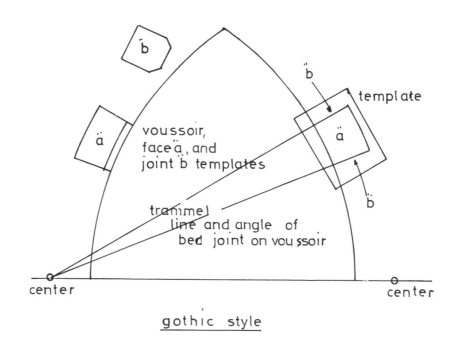

b̈

b̈

template

ä

voussoir, face ä, and joint b̈ templates

ä

b̈

trammel line and angle of bed joint on voussoir

center

center

gothic style

Figure 37

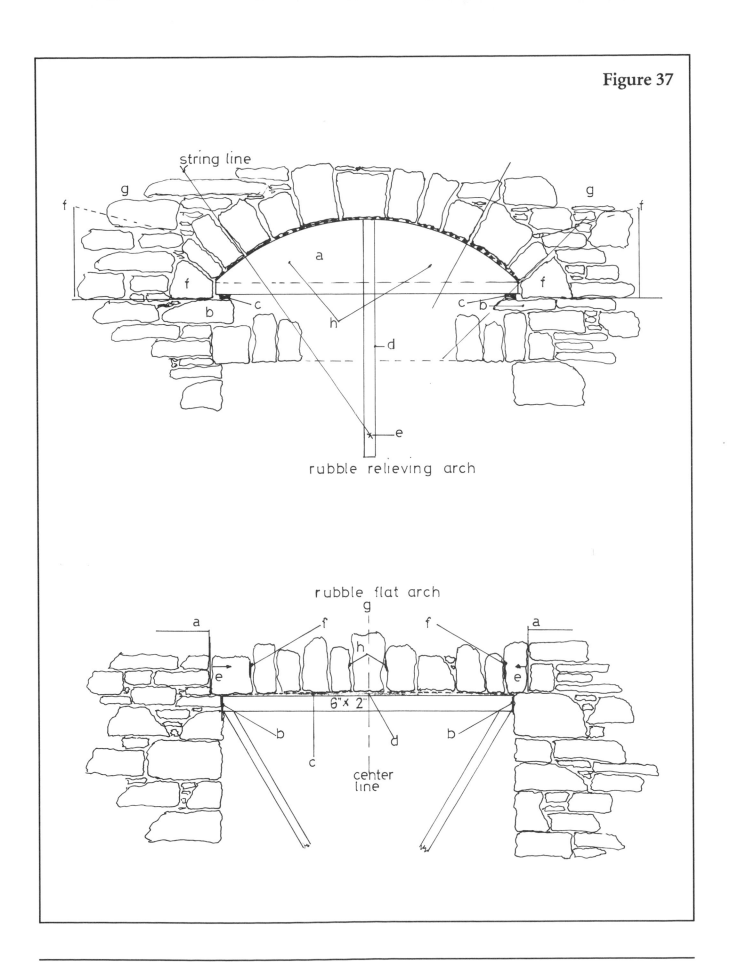

string line

g — f

g — f

a

f — b — c — h

c — b — f

d

e

rubble relieving arch

rubble flat arch

a — f — g — f — a

e — h — e

6" × 2"

b — c — d — b

center
line

Figure 38

flat and relieving rubble arch

Figure 39

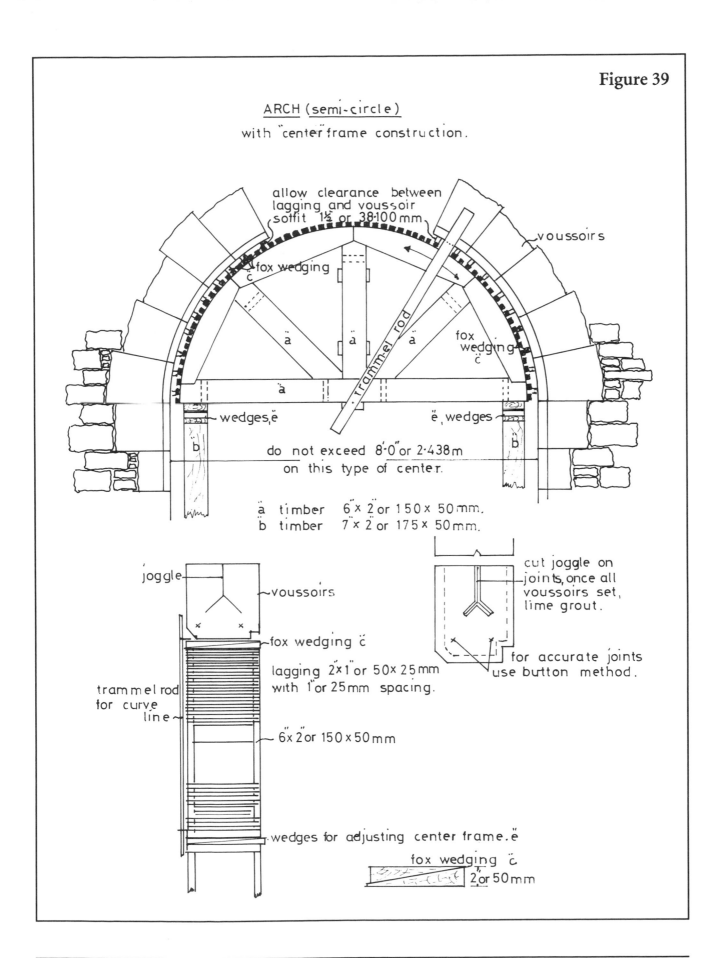

ARCH (semi-circle)
with "center" frame construction.

allow clearance between
lagging and voussoir
soffit 1½ or 38·100 mm

voussoirs

fox wedging
c

trammel rod

fox
wedging
c

a   a   a

a

wedges, ë   ë, wedges

b   b

do not exceed 8'·0" or 2·438m
on this type of center.

a timber   6"× 2" or 150× 50 mm.
b timber   7"× 2" or 175× 50 mm.

joggle
voussoirs

cut joggle on
joints, once all
voussoirs set,
lime grout.

fox wedging c

lagging 2"×1" or 50× 25 mm
with 1" or 25mm spacing.

for accurate joints
use button method.

trammel rod
for curve
line

6"x 2" or 150 x 50 mm

wedges for adjusting center frame. ë

fox wedging c

2" or 50mm

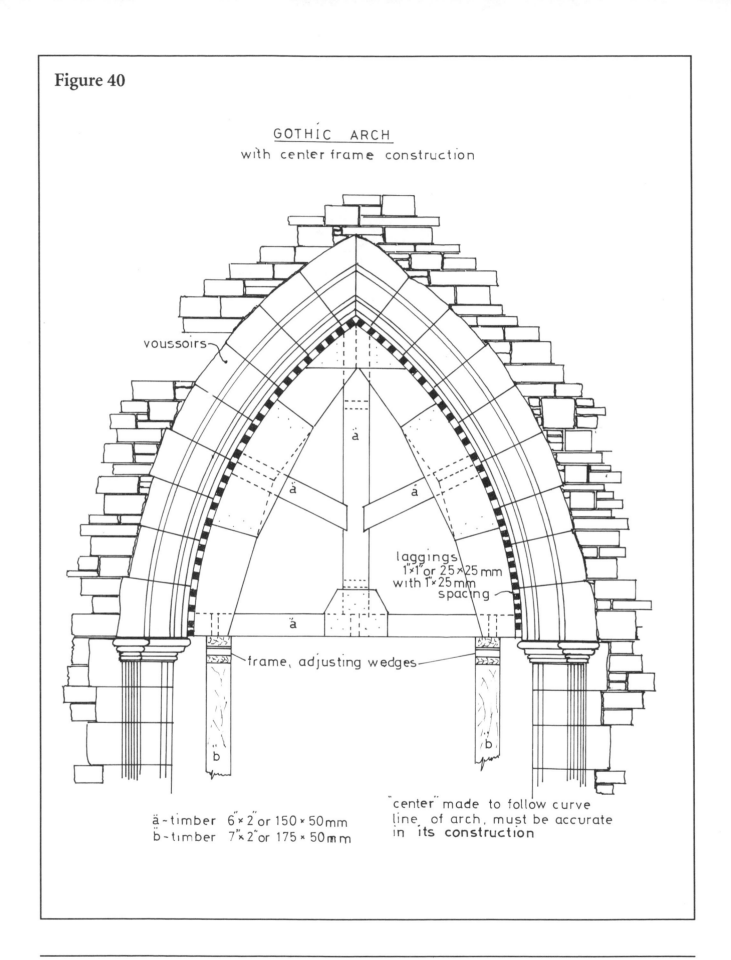

Figure 40

GOTHIC ARCH
with center frame construction

voussoirs

ä

ä

ä

ä

laggings
1"×1" or 25×25 mm
with 1"×25 mm
spacing

frame, adjusting wedges

b

b

ä—timber   6"×2" or 150×50mm
b—timber   7"×2" or 175×50mm

"center" made to follow curve
line of arch, must be accurate
in its construction

Figure 41

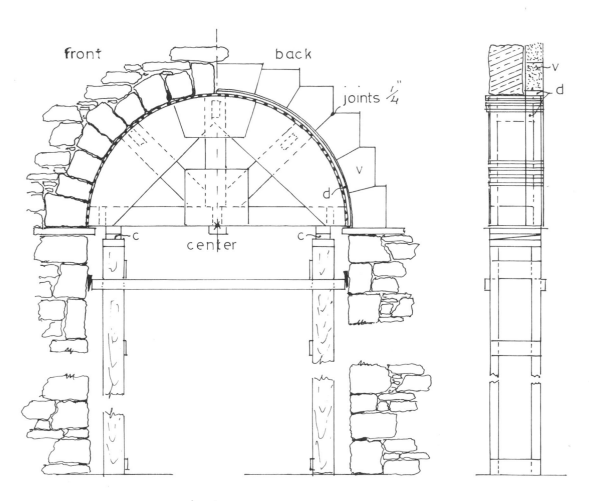

front   back

joints ¼"

v

d

center

c   c

<u>semi-circular arch</u>
<u>rubble front   precast voussoirs back</u>

fox wedges c

voussoirs
use ¼"x½"
ply for spacers

mortar

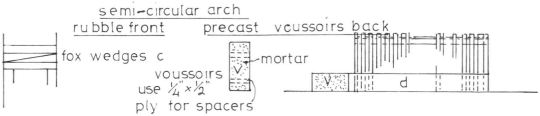

d

to make voussoirs, lay frame on floor,
cast in cement, using ¼" ply as joints.

d — use 3⁄16" plywood strips in layers, nailed to laggings
to required depth

Figure 42

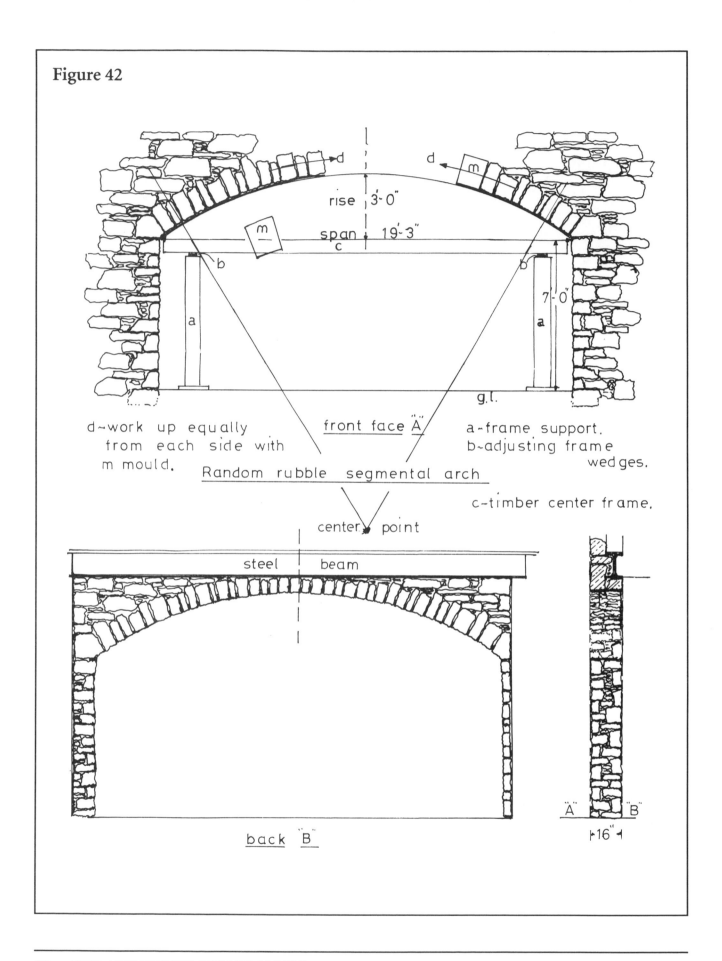

rise 3'-0"

span 19'-3"

7'-0"

g.l.

d~work up equally from each side with m mould.

front face "A"

a-frame support.
b~adjusting frame wedges.

Random rubble segmental arch

c~timber center frame.

center point

steel beam

back "B"

"A" "B"

|-16"-|

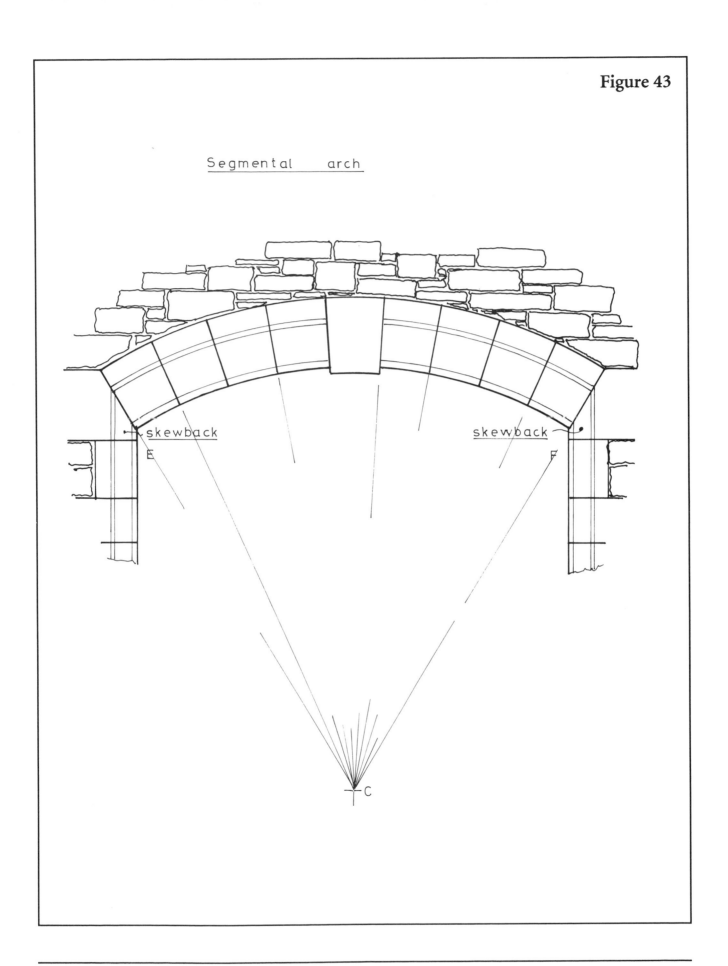

Segmental arch

skewback

skewback

Figure 43

Figure 43 - A

rybits
or.
quoins

elevation

elevation

plan

window

plan

door opening

do not use button for bedding on
your rybits, beds are usually
rough, and uneven, also vary
in thickness.
center frame required for
door opening arches.

mortar

rybit or quoin
use bedding rod only, with
rawhide hammer to adjust.

rybits
or.
quoins

elevation

plan

door opening

Building a Traditional "Sow" or Hunchbacked Bridge

For the following discussion, refer to Figures 44-48. To build a "sow" or hunchbacked bridge, you must first set out the lines for the foundations and the position of the bridge, remembering your 3-inch scarcement below ground level. Having got your wall lines and pegs set, excavate the founds, removing the topsoil until you reach the hard subsoil. You do not need to go deep — say, about a foot below the subsoil. If the banking at the sides slopes, use a step foundation as shown in Figure 2 on page 20; each step must be level.

Build up your foundations with scarcement to about 3 inches below ground level. Then start by building your walls, including the butts, as in building random rubble, up to the springer line on the arch, leaving risers going through ready for the next lift. You build up to this section with solid stonework, grouting each layer with your building mix. Do not throw infill stonework in, but build it as in building a double-faced wall. When all this is solid, you should have your timber center frame ready as described in the section on arch construction. Set up the frame, making sure you fit in your frame, adjusting with fox wedges. This is to make it easy to remove the frame once the arch is complete. Allow a 1-inch clearance between the frame and the soffit of the stonework for a bed of soft sand, dampened and spread over tar paper, which has been laid on the center frame. This allows the snouts of the stone to press into the sand and gives an even surface to the stonework.

Simple bridge building in random rubble.

This also allows you, once the voussoirs are in position, to pour dry sand in between each joint, to an average depth of $1/2$ inch. This keeps the grout from running through, keeps the soffit clean, and makes it easy to point, as the dry sand runs out on removal of the frame.

Start by building up your outside faces first, from your springing line level. Once this is complete, start the outside voussoirs, working up from each side until you reach the center point. Now do the infill arch stones, tying over and setting your stone tipped at the joints only with your mortar mix. This is to allow clearance for your grout. If possible, try to work your infill arch over completely before doing any grouting.

Grout using a mixture of 3 parts sharp sand, 1 part lime, and 1 part cement. Fill the total area up to the top of the stone arches. Then throw in gravel, allowing it to go down into the joints; the pieces of gravel will act as wedges.

Next, build your infill or spandril behind the arch stones as shown in Figure 45. This acts as a saver, meaning you build hard against the back of the arch stones, overlapping each stone. This is to save your arch carrying all the weight. Grout each layer as you build up. Do not use concrete. Build up until you are about 2 inches below the bottom of the slabs you will use as the final finish.

Once you have leveled off to slab finish, build the die walls, as in building double-faced random rubble, about 1 foot thick, up to the required height, finished ready for the coping. (See coping detail, Figure 17 on page 34.)

All small stone bridges are made as in building an arch, making sure that the spandril, where the load of the arch is applied, is solid and well-built.

This bridge has a crazy paving finish.

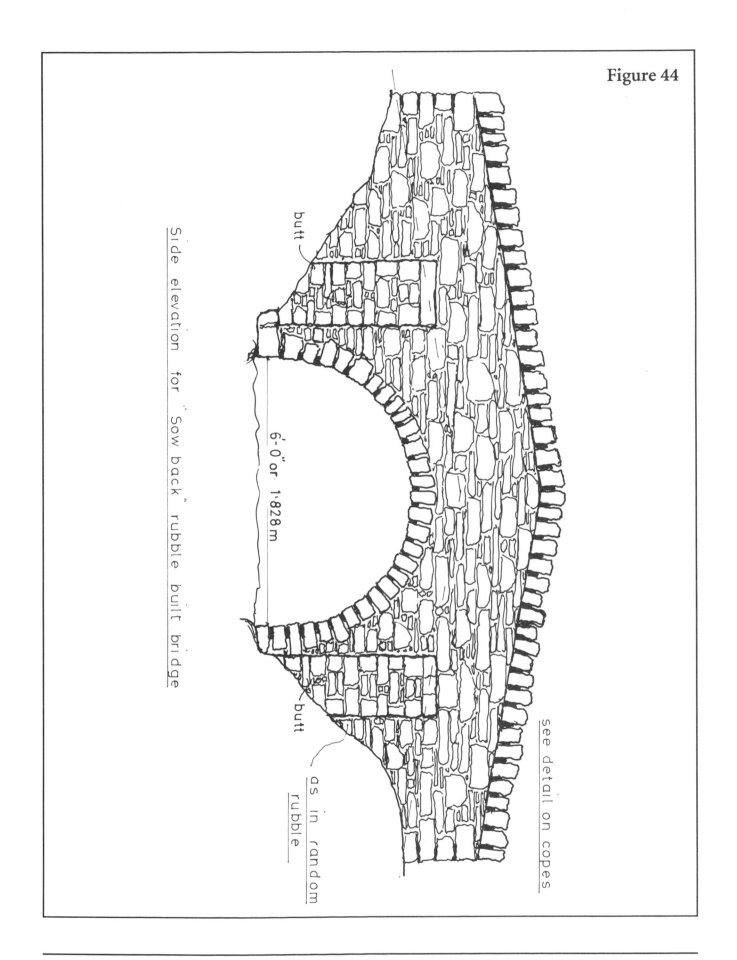

Figure 44

Side elevation for "Sow back" rubble built bridge

butt

butt

6'-0" or 1·828m

see detail on copes

as in random rubble

Building a Traditional "Sow" or Hunchbacked Bridge   87

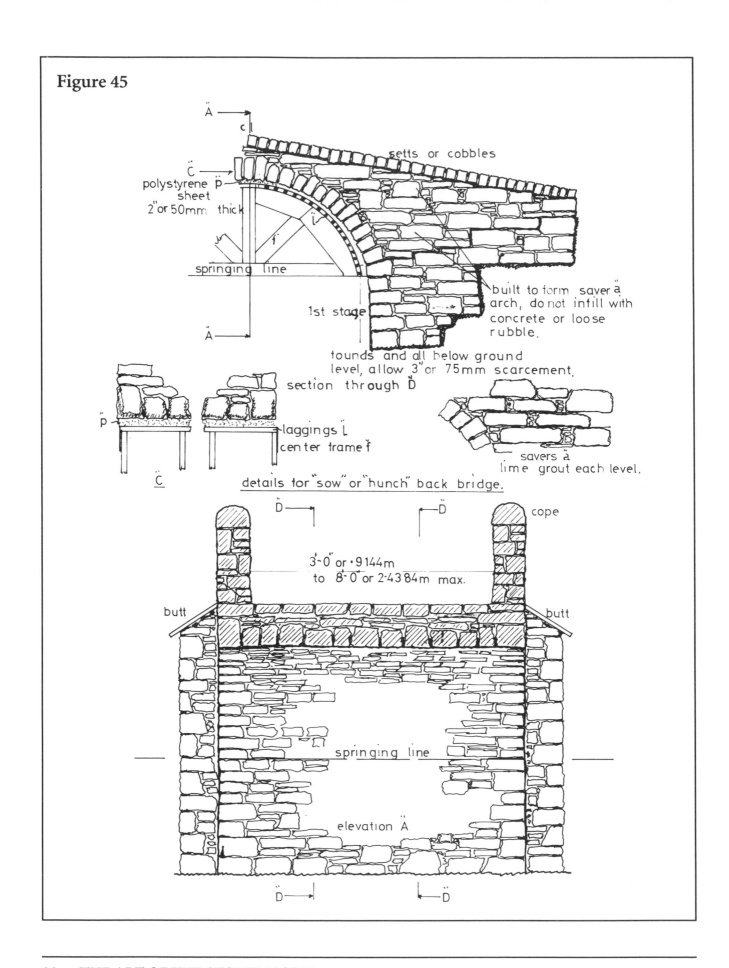

Figure 45

setts or cobbles

A

C
polystyrene p
sheet
2" or 50mm thick

springing line

1st stage

A

built to form saver ä
arch, do not infill with
concrete or loose
rubble.

tounds and all below ground
level, allow 3" or 75mm scarcement.

section through D

p
laggings L
center frame f

C

savers ä
lime grout each level.

details for "sow" or "hunch" back bridge.

D   D   cope

3'-0" or ·9144m
to 8'-0" or 2·4384m max.

butt   butt

springing line

elevation A

D   D

Figure 46

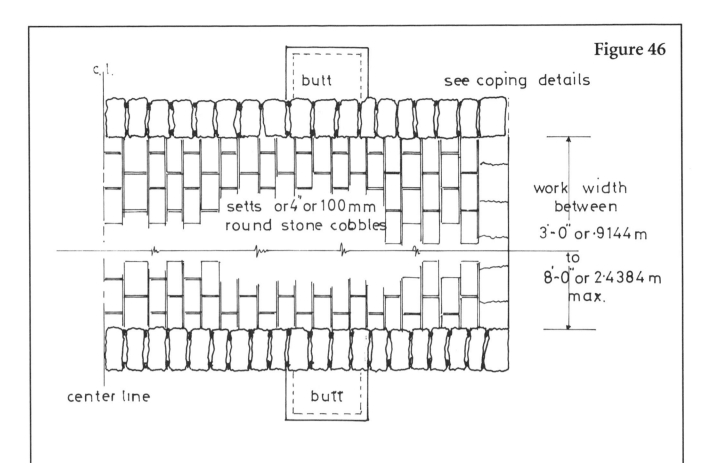

c.l.

butt   see coping details

setts or 4" or 100 mm
round stone cobbles

work width
between
3'-0" or ·9144 m
to
8'-0" or 2·4384 m
max.

center line   butt

plan of "sow" or "hunch" back bridge.

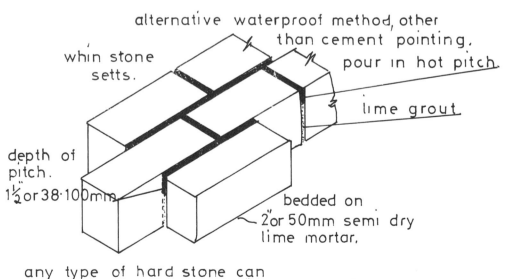

alternative waterproof method, other
than cement pointing.
pour in hot pitch.

whin stone
setts.

lime grout

depth of
pitch.
1½ or 38·100mm

bedded on
2" or 50mm semi dry
lime mortar.

any type of hard stone can
be used, not more than 7" or 0·1778 m wide.
not less than 6" or 0·1524 m deep.

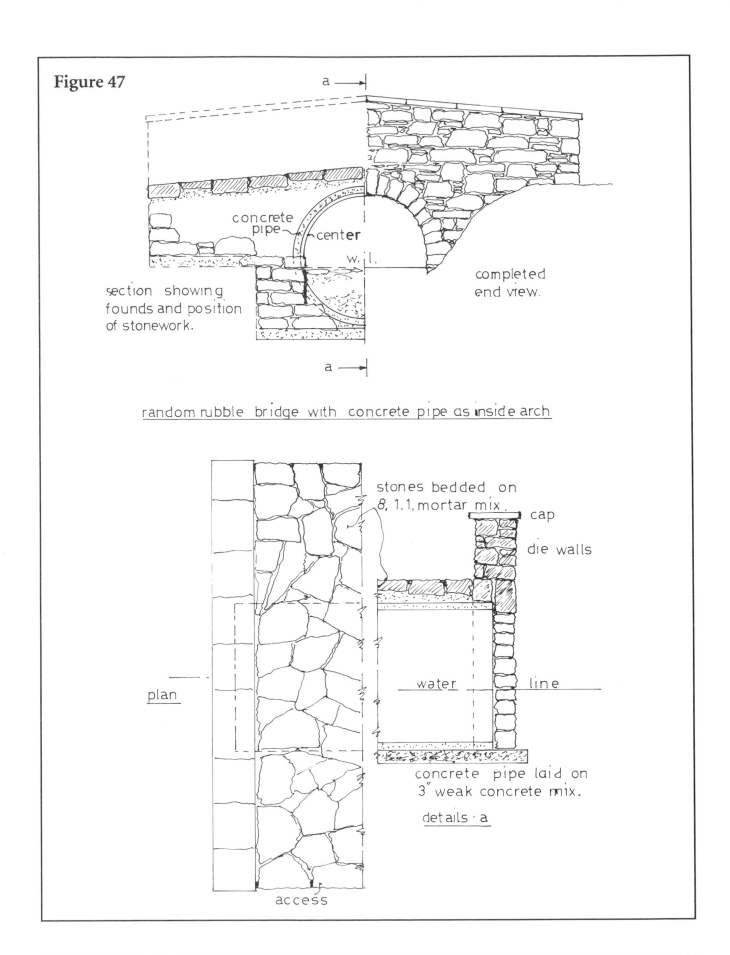

Figure 47

a →

concrete
pipe
center
w. l.

section showing
founds and position
of stonework.

completed
end view.

random rubble bridge with concrete pipe as inside arch

a →

stones bedded on
8, 1.1, mortar mix.

cap

die walls

plan

water   line

concrete pipe laid on
3" weak concrete mix.

details · a

access

Figure 48

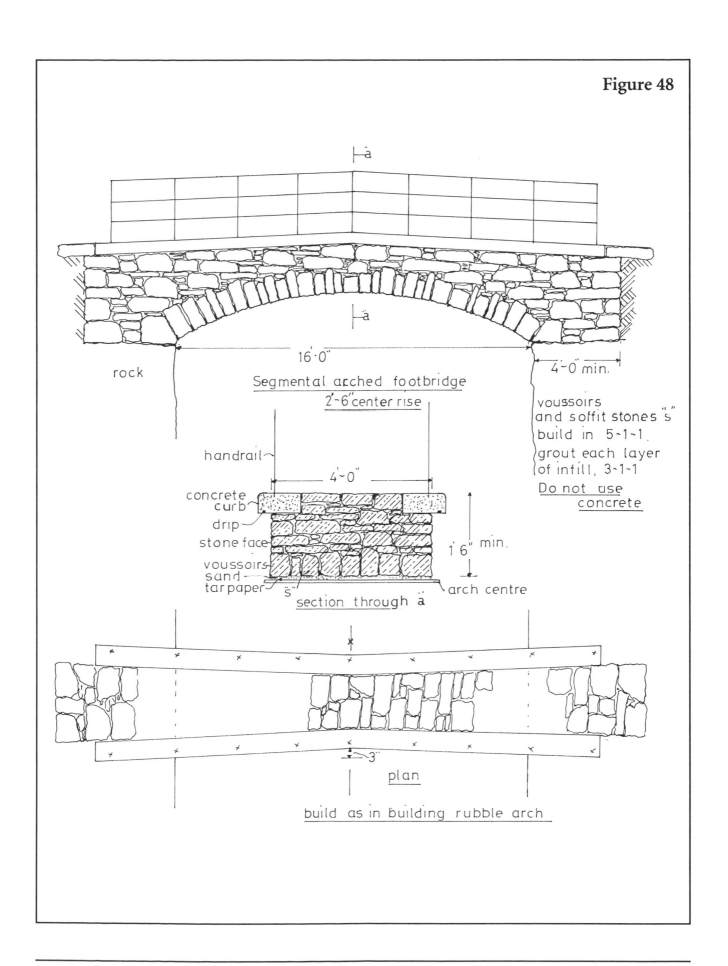

rock

16'·0"

4~0' min.

Segmental arched footbridge
2'-6" center rise

voussoirs
and soffit stones "s"
build in 5~1~1
grout each layer
of infill, 3~1~1
Do not use
concrete

handrail~

4'~0"

concrete
curb
drip
stone face
voussoirs
sand
tar paper   s"   arch centre

1' 6" min.

section through a

plan

~3"

build as in building rubble arch

Restoration Work

I find restoration work — trying to recreate the original work — the most satisfying I have done. We often know very little of the original methods of construction, as nothing was recorded. Any information was usually handed down through the generations, through father to son, by verbal instruction only. Each part was secretly guarded from anyone outside the family, with each stonemason having his own mark.

The beautiful abbeys and cathedrals found throughout the United Kingdom were built between the 12th and 14th centuries. With all the skill and knowledge we have at present, we could never repeat the magnificence of this building period.

During this period of cathedral building, the Church was the dominant force. It employed and controlled the stonemasons as well as the other craftsmen involved in construction. In 1120 Bishop Jocelyne formed the Guild of Stone-masons, today referred to as the Free Masons; their purpose was to advance and protect the art of building and working with stone.

After this period the Crown took over from the Church, and a great castle and palace building program began. The Crown would conscript stonemasons, carpenters, stoneworkers, and laborers from each town and village. Laborers were expected to pick up their tools and start walking to the designated contract. They did so.

Once the country had settled down to a more peaceful way of life, and the people had greater wealth, the castle building era slowed, and ruling landlords established vast estates. Large mansion houses were built throughout Great Britain, with each landowner trying to create a masterpiece of architecture, building, plastering, and decor. Examples of much of this work are still in good condition and lived in to the present day. A number of the old buildings and estates are looked after and maintained by the National Trust.

In working on the restoration of these old buildings, you are working back into the history of the building, learning the secrets of those methods of construction and the mortar mixes used. You will find buildings that have been standing 300 to 400 years are still in good condition, as far as their construction is concerned.

Up to 1220, cutting a stone was done with an axe, as you can tell by the light relief work of the stones' carvings. After 1220 the chisel (as we know it) came into use, and from that period on the cutting mason became a highly skilled craftsman. Many cut elaborate moulding work. At one time this was a separate branch of the stonemason trade.

To a stonemason, old walls can tell many secrets. You can tell their age, and roughly when they were built by the mortar that was used, by the stonework, and especially by the mortar infills in the heart of the walls. You can even tell the mood of the stonemason who was working on random rubble buildings. As you work in the footsteps of the original stonemason who carried out this work, you are lifting the same stones he lifted when he built this wall hundreds of years ago. If, as you are working, you think deeply on any problem you might have, so might he have done. With a little thought as to what he might have been like, what kind of man he was, you will

---

Preservation — To help keep in a sound state, to maintain in good condition what is already there.

Restoration — To reestablish what was originally there, to repair and make strong again.

This wall was restored in 1985 to conform with the 13th century rubble work. The pointing was done with a piece of wood to give it an antique finish. The wall thickness at this point is 6'6".

find that long-ago stonemason might solve your problem. Nothing is impossible in building stone, if only you know how. It has all been done before.

Restoration work requires knowledge of traditional building methods, patience (it is time-consuming work), caution (it can be dangerous), and above all, common sense. You do not thump the wall with a hammer if you have fifty tons of stonework hanging above you. Restoration work can be very difficult, especially on very old buildings that have been allowed to deteriorate through the years. Often, plant growth has penetrated the walls, and, worst of all, large tree roots may be growing out of the most awkward sections of the building.

I will explain, through text and illustrations, some of the methods I have used in restoration work. No two buildings have the same problems.

In doing this work, always wear boots with a safety toe cap, as any stone dropping usually makes for your feet. Keep the area around the work clean and tidy. Do not have too many tools and materials blocking your way for a quick exit, should anything above you give way. Do not wear loose clothing, rings, or other items that might catch onto the surrounding work.

I do not wear a safety hard hat, though I know I should. I usually feel dust particles hitting my cap prior to the stone dropping, giving me the chance to jump clear. I prefer dust hitting my cap to a five-ton lump of masonry bouncing off a hard hat.

Invasion of Vegetation

Study the work you are undertaking very carefully. I have found through experience that most restoration work involves random rubble, with walls from 3 to 6 feet thick, built in lime mortar, with defective joints and plenty of vegetation. This vegetation holds sections of the wall together. Tree roots may go well down into the wall, dislodging the stonework or forcing it to bulge.

Staircase of Mugdock Castle before restoration.

The easiest method for this type of restoration is to cut out a section of the stonework, removing all growth. Then rebuild the section to conform with the existing part. Go on to the next section, continuing this method until all growth and roots are removed. The area being restored should conform to the existing stonework. Never try to pull a root or tree out, as this simple action will bring down the wall.

Alternatives to Shoring

I do not recommend using timber shoring or jacks if the stonework has been hanging for a number of years. Any movement in fixing or tightening a jack could result in the whole wall coming down on your head!

If I feel a section may drop, I use a *wet wedge*, having used this method many times with success. Again, it requires common sense and a thorough knowledge of stonework. The idea of the wet wedge is very simple. An ordinary white pine wedge is soaked in water overnight, allowing it to swell. Then it is pressed into the open upright joints, not into the bottom bed joints of the stonework. Do not hammer it in, as the wedge is wet and already swollen, giving it a tight fit. Using a dry wedge, press the wet wedge in tight, as the moisture causes the wedges to swell. If you used a dry wedge, the moisture from the surrounding stone would cause the wedge to swell, either fracturing the stonework or forcing the area around the wedge to collapse.

Removing Old Cement Pointing

A problem I find in some restoration work is that the stonework has a hard cement-rubbed pointing on top of the old lime-built joints. The cement pointing, done a number of years later, was defective and had allowed water penetration, and owing to the hard surface the water could not get out. With severe frosts throughout the years, considerable internal damage resulted. In some

Staircase of Mugdock Castle after restoration.

Traditional masonry showing stone relieving arch over timber lintol. This section of the 15th century building was originally the basement of the great hall above.

Traditional masonry showing position of floor joist rests on this rough squared rubble building built in 1840.

areas the lime had washed into the heart of the wall and, with more frost, caused the wall to bulge in sections. This is one reason I ban the use of strong cement mortars in building and pointing stonework. Walls that are lime-pointed have been standing for many years.

To restore such a wall, cut away from any stone the cement pointing surrounding it. You will find it damp. Also, remove a stone, such as a quoin, that has been bedded on a strong cement mortar, and you will find that the bed can be removed easily. It will resemble a slate, with very little adhesion to either surface of the stones.

Grouting Stonework

In grouting stonework, I use a lime-based grout, not a strong cement grout as some recommend. Open up a wall that has been cement-grouted, and you will find little adhesion to the surrounding material. Examine any cement pointing on stone, and you will see hairline cracks between the pointing and the stone, allowing for penetration of water.

The lime grout, like any lime pointing, seems to be absorbed into the stone, and any movement within causes the two to move together. The mixture, when dried out, should be as in old lime-built buildings — tough and flexible, absorbing any moisture penetration, and yet able to dry out very quickly. This method of building gives the walls a chance to breathe.

I also do not believe in pouring gallons of water down through the heart of the wall prior to grouting. Water pushes all the sediment in the heart of the wall into pockets, and being wet, will not absorb your grout. I only dampen the area and pour in my lime grout. This goes through all the old lime and gives a good, solid structure, as it adheres to and is absorbed into the stone.

Correcting Patching

Another problem created by cement rendering is patching already done to defective stonework (Figure 49). This method of coloring cement and dusting with sand to try to conform

with the existing material does more harm than good. The cement is a hard, impervious material with no "give" to it. The stonework absorbs moisture, so frost can damage the patch, causing hairline cracks, which allow further water penetration and moisture to gather at the back of your cement patch. This can cause stone rot, as the stone is not getting the chance to dry out. Stone rot requires the skill of a stone restorer to rectify it. Do not think you are doing any stonework a good turn by coating it with cement.

On no account should sand blasting, acid washing, or any other abrasive method be used in cleaning stonework. These methods remove the natural impervious surface of the stone, allowing the polluted atmosphere to penetrate into the stone and hasten the decaying process.

Restoring an Arch

Examples of some of the restoration work I have done can be seen in Figures 50 and 51. The illustrations show a missing arch rib, with the spandril arch dropping in, ready to collapse. This damage was caused by large stones dropping off the wall head of the tower. The arch formed the third floor and ceiling of the second floor within the tower. Further damage was caused by tree roots and growth from the springing line of the arch.

First, to make the area safe to work on, I removed the defective stonework and cut out the tree roots.

A timber arch center was required, and owing to the restricted area for working and the potential for danger in building the center underneath this stonework, the center was built out in the courtyard and made up in kit form. It was then transported in pieces and erected in position where the new rib was to be built.

With the center in position, I blocked up and wedged between the lagging on top of the center and the soffit of the sagging spandril arch. I then felt the structure was safe to work on. I started working my way down on top of the spandril arch, removing all defective stonework and growth right down to the springing line. This was all cleared on one side of the arch only. Next, I wedged up the spandril arch stones into their original positions, cut and fitted new stone keys to lock the arch stones in position, and poured over a lime grout to seal all joints. I proceeded to rebuild the spandril arch infill exactly the same way as it came out, placing the stones as in forming a saver lintol, grouting each layer as I worked my way up. Completing the one side, I repeated the same procedure on the opposite side.

The spandril arch could not be carried over and sealed, as I had to have an access point to bring through scaffolding, stones, and other material, as shown on the drawing. The lifting gear used for bringing up the spandril arch stones, illustrated in Figure 52, was easy to erect and handle.

With the spandril arch and the infill dry, I was ready for the next stage. I removed the wedges and the blocks, which held the spandril arch in position until it had set. With the *voussoirs* ready for the ribbed arch, I lifted each one up through the opening and laid it on top of the timber center. They were then slid into position using fox wedges and a trammel to get the true line of the curve. I carried this out working up from each side equally, until the ribbed arch was complete. Before setting each voussoir, I cut a small hole in each joint face, spread my lime bed, and set buttons for accurate jointing. Into the holes I had cut, I placed a small pebble to act as a key between each stone. This prevents any lateral movement and gives a little extra strength at the joints. When the ribbed arch was complete, the joint between the top of the ribbed arch and the soffit of the spandril arch was pointed with a lime mix.

The rest of the spandril arch stones were brought through the opening, plus a new key stone for the next ribbed arch, as this and four spandril stones were badly fractured. The four fractured spandril stones were sagging badly. I could not build up a scaffold to the underside, because in working underneath the structure and in removing the stones, the whole lot could have dropped.

I cleaned the top surface of the damaged stones and inserted fine, strong wire through the cracks to make a clamp. This allowed me to hold and lift the stone back into a safe position.

Once the damaged stones were removed, I set the new stones in position, sealing all over with a lime-based grout. The main ribbed arch was then complete. I slackened off the holding wedges of the timber center, allowing the fox wedges to be removed easily. Then I moved the complete center to the next rib, the one with the badly fractured key stone. With the center in position, I tightly wedged each stone in the ribbed stone, including the key stone. To make sure the wedges were tight, I threw some water over the wedges, causing them to swell and tighten further. I then proceeded to cut out the keystone. Once it was removed, I cut a *joggle* (a joint made between stone surfaces by cutting a notch in one and cutting the other to fit the notch; this prevents any movement between the stones joined) on each face joint of the stones, formed my bed joints, and inserted the new key stone, wedging it into the original position, then grouted the joggle joint.

The ribbed arches were complete, so I dismantled the scaffold, removed everything not needed to the floor below, sealed the opening with spandril stones laid aside for this purpose, and sealed all with a lime grout.

Restoring a Window

The restoration pictured in Figure 53 required careful study; a careless move could have brought the building down. The damage was caused by weather deterioration and growth of vegetation, with a little help from vandals.

First, I gave it a dry clean with a fine brush. Then I sprayed the area with a watered-down glue, especially where the old lime mortar was exposed. This spraying was to harden the exposed lime surfaces and prevent further deterioration. I gathered together the door quoins, or rybits, so everything was ready.

Traditional masonry construction of squared random with infill snecks. It was built around 1820 and now awaits a restoration mason.

A ribbed arch at Mugdock Castle tower, which was rebuilt in the 1630s. At the level shown by the corbel projection, the wall thickness is 5'6"; at the top of the tower it is 4'6".

I then proceeded very carefully to set each stone and work it into position, avoiding the use of a rawhide hammer. I built up to the underside of the existing lintol, leaving ¹/₄ inch between the top bed of stone and the soffit of the lintol. Using my fine tamper and a damp, lime-based mortar, I tamped this mixture hard in, avoiding the use of slate wedges, which might have disturbed the stonework above.

With the door section dried out, I began work on the window. I cut down and removed defective stonework below the window sill level and rebuilt this, forming new seats for the sill, then set the sill using hollow bedding. I gathered together the window quoins, built them into position, and built the surrounding stonework up to the height of the lintol.

The lintol was broken, which further complicated the setting process. I made a frame, set this in position with a layer of sand on top, placed the lintol on top of the sand, and built in from each end to tighten, as you would do with a flat arch. I cut stones to form the *saver arch* between the top of the lintol and the existing stonework. The weight of the saver arch was applied to both ends of the lintol, and the center of the bed joint was filled with a weak lime mix. The purpose of this was to prevent any weight being applied to the center of the lintol.

Once this elevation had dried out and was completed, I gradually removed the wet wedges I had used for shoring it, and tamp-pointed each joint and the surrounding stonework.

The front elevation was now safe from collapse, but the inside or back elevation, which was composed of random rubble-built walls 4'6" thick, was in very poor condition. I made good the jambs of the window, up to the lintol height. The original had been a timber lintol, which had rotted away. I decided to cover the opening with a flat arch. This eliminates heavy lifting in a position that is dangerous and difficult to work around.

I set up a frame and mixed a weak mixture of lime and sand, the lime acting as a binder. This mix was tamped hard on top of the frame to give the actual line of the flat arch, allowing a ¹/₈-inch rise per foot to the center line of the arch.

Having hammer-cut the arch stones, I set these stones, working up from each side on the sand, bedding only at the joint edges. This action allows the stones to sit in their actual positions and allows a free run for grout to go in between the joints. After setting all the stones in position right to the back of the front lintol, I fitted the key stones in the center. The arch was now formed. Next, I dropped dry sand in between the open joints of the arch stones, filling up to about ¹/₂-inch from the bottom. Once the sand was in all over, I grouted everything with a mixture of 3 parts sand, 1 part cement, and 1 part lime up to the top of the stones. Then I threw some gravel into each joint, pressing down gently. The pieces of gravel act as wedges, preventing any movement. Once this had dried out (at least twenty-four hours), I built up in random rubble to conform with the existing work, sealing the opening above the arch. Next, I cleaned out the lime and sand mix used for the arch, removed the frame, and brushed and cleaned out the sand put in prior to grouting. This gives a good joint for pointing; the sand prevents the grout from running through and staining the stonework.

The stonework was then cleaned and pointed to conform with the existing structure.

A Leaning Chimney

A leaning chimney is a common problem encountered in restoration of old buildings, especially if the chimney stack rises to a good height above the apex of the roof. The number of vents in the stack also affect its structure. Chimney stacks were normally built of dressed cut stone to match with the exposed wall face supporting the chimney. The internal face (the part inside the building) would be random rubble construction, the thickness being 6 to 7 inches. To save time and stone, most such chimneys were built on *cant* (off bed or on edge). The weakest point in the stack is where the dressed stone rests on the internal rubble-built wall. You will find that the stones that are on cant have given way, taking the internal vent (stone) bridges with it. Bridges are the stone walls usually 6 inches thick. They separate each vent and are very rarely tied over. My method is as shown (Figure 54) for realigning chimneys that lean in or out. If inserting new vent linings when rebuilding, do not use strong cement; surround the new vent linings with lime-based mortar.

A Corbie Stepped Gable and a Random Rubble Gable

A corbie stepped gable (Figure 55) is built by constructing a number of steps formed of hard cut stone squared, the top surface being slightly weathered. This is also known as "crow step gabling." The method of restoration involves setting off the first step, and, using lines, to line

down from an outrigger. When setting and pointing corbies, do not flush point at "p", as this will allow water to drip off and not run back into the mortar bed of stone.

Restoration and rebuilding of a random rubble gable is pictured in Figure 56. Having leveled off the wall head, leaving some risers going through, you set up your scale couple. This is made up of the roof rafters and collar beams you are going to use in the roof construction, and will give a true line to your roof. Set this up plumb as shown on the drawing. With your outrigger, drop a line with a plumb bob at point "b", then work this line backward or forward until the point of the plumb touches line "a". Having plumbed down, mark the line at "b", then fit lines marked "d" from point "b" to point "a". Build up as in random rubble building, using line "d" as the plumbing.

The author with his son John at the completion of a section of restoration at Mugdock Castle, Scotland, 1985. The stonework was built to conform with the original 14th century work.

This is the window referred to in Figure 49 on restoration. It was restored in 1984 to conform to the existing 17th century building. Here it awaits pointing. Note the saver lintol above the flat stone lintol.

Figure 49

RANDOM RUBBLE (rebuilding patches)

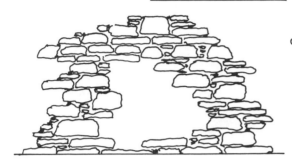

doing repair patches in random rubble.

Using large stones with strong cement mortar, forms a hard patch with weak surrounds, usually gives way, pulling down surrounding rubble. ──────►

wrong

line

build up in lifts of 18" or 460mm, using your lines, and stones to conform with existing rubble; use lime mortar mix.

patch complete, should conform with existing.

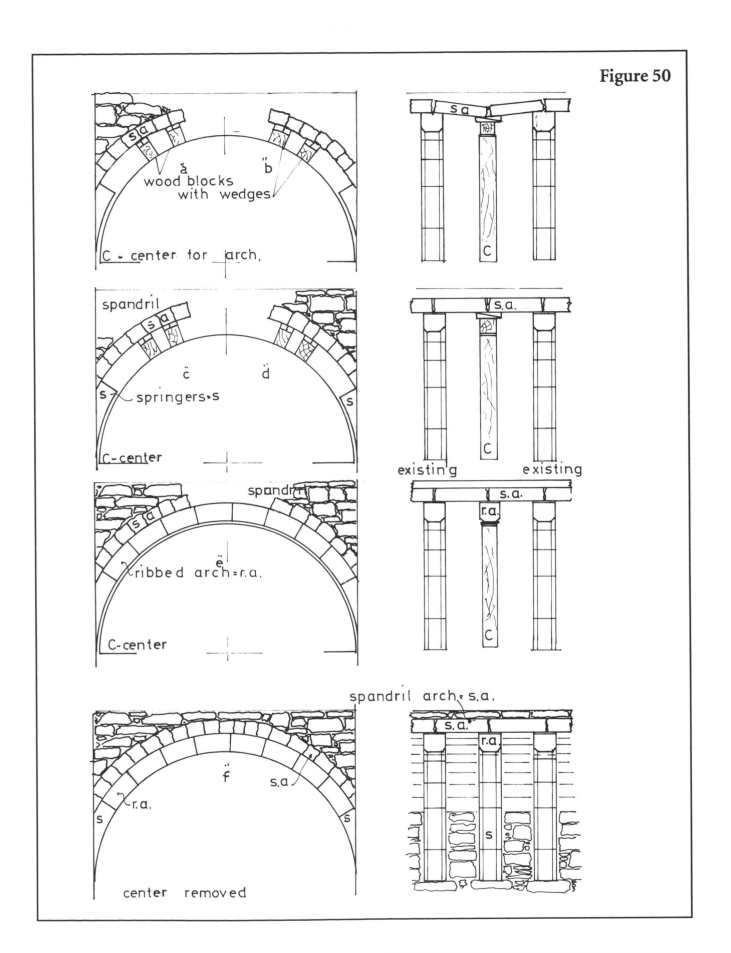

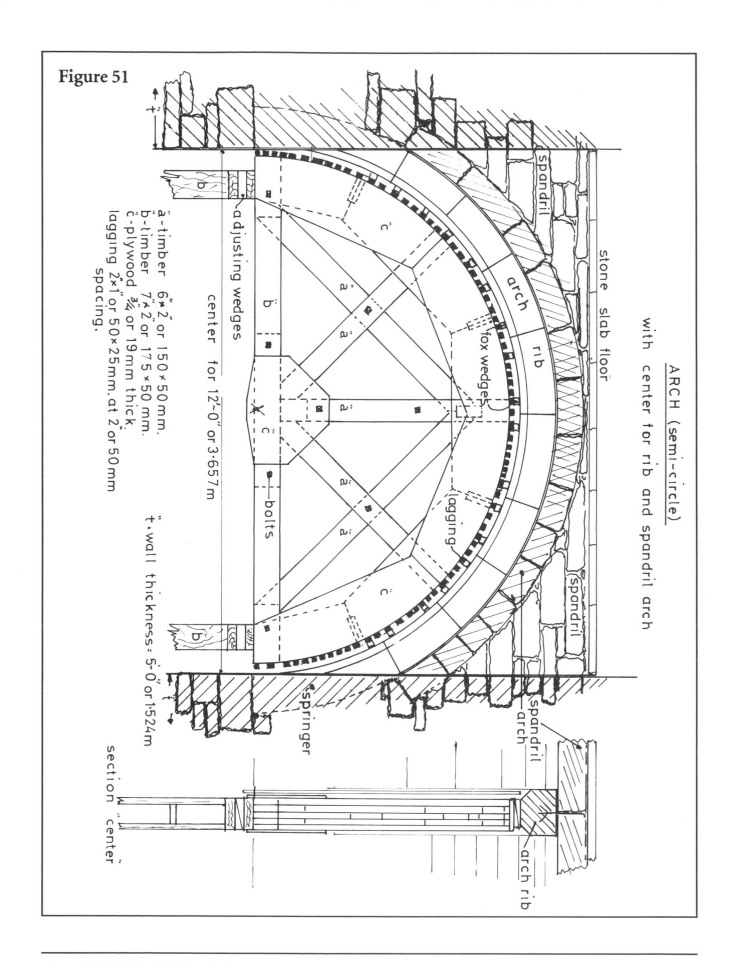

Figure 51

ARCH (semi-circle)

with center for rib and spandril arch

spandril

arch

rib

fox wedges

lagging

stone slab floor

spandril

spandril
arch

arch rib

springer

section center

adjusting wedges

center for 12'-0" or 3·657m

bolts

t· wall thickness = 5'-0" or 1·524m

a-timber 6"×2" or 150×50mm.
b-timber 7"×2" or 175×50 mm.
c-plywood ¾" or 19mm thick.
lagging 2"×1" or 50×25mm, at 2" or 50mm
        spacing.

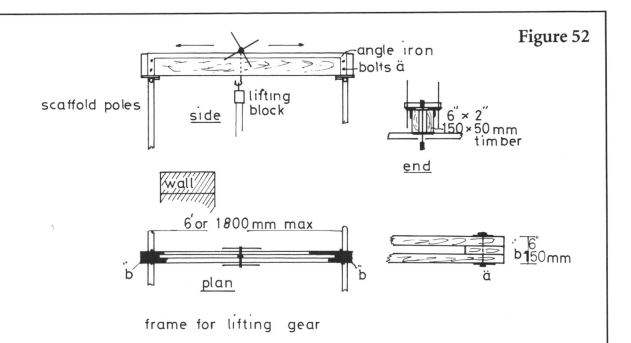

Figure 52

side

scaffold poles   lifting block

angle iron bolts ä

6" × 2" 150 × 50 mm timber

end

wall

6' or 1800 mm max

plan

b   b   ä

6" 150mm

frame for lifting gear

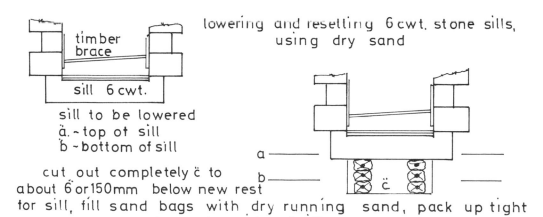

lowering and resetting 6 cwt. stone sills, using dry sand

timber brace

sill 6 cwt.

sill to be lowered
ä - top of sill
b - bottom of sill

a

b

cut out completely c̈ to
about 6' or 150mm below new rest
for sill, fill sand bags with dry running sand, pack up tight
to under-side of sill.

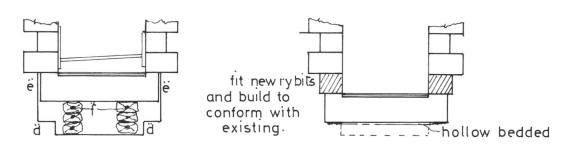

fit new rybits
and build to
conform with
existing.

hollow bedded

once this is tight, cut down to form new
sill rest-d̈,ë-cut around to clear sill, once clear, tap sill to slacken
and rest on sandbags. open sandbags and allow sand to run out
slowly†, making sure sill drops slowly on to its new rest.
simple hydraulics using sand.

Figure 53

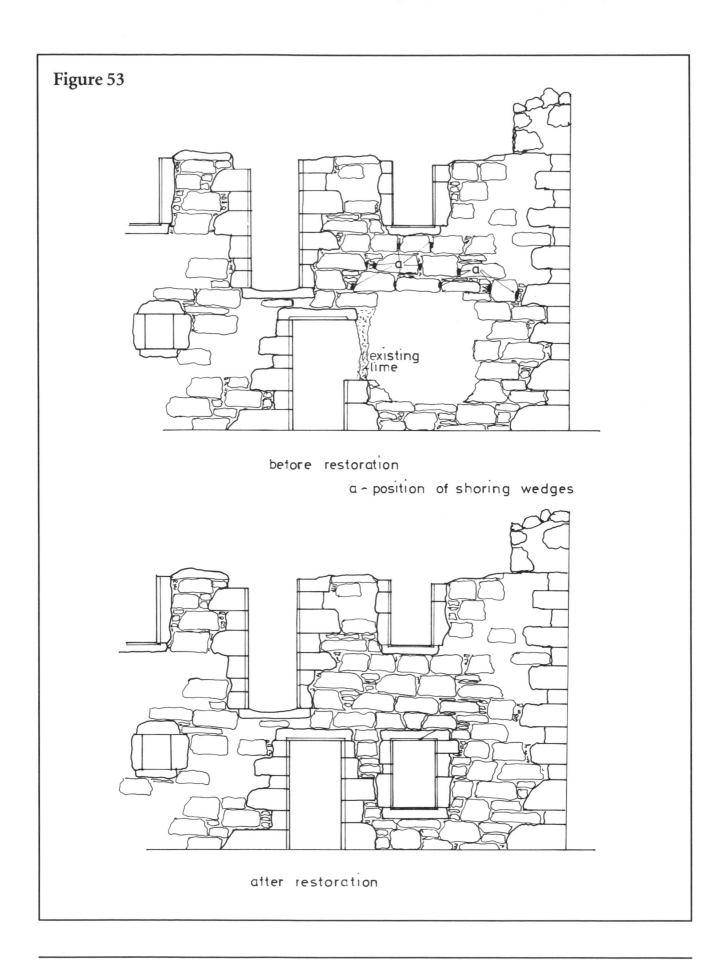

before restoration

a - position of shoring wedges

existing lime

after restoration

108   THE ART OF THE STONEMASON

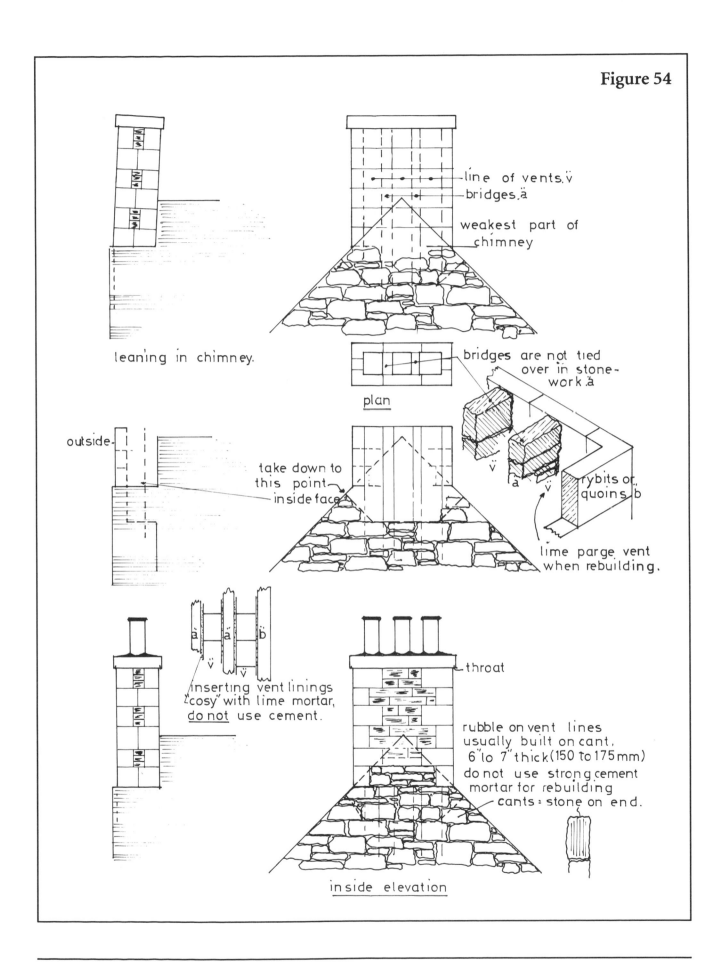

line of vents. v̈
bridges. ä
weakest part of chimney

leaning in chimney.

plan

bridges are not tied over in stone-work. ä

outside

take down to this point inside face

rybits or quoins. b

lime parge vent when rebuilding.

inserting vent linings "cosy" with lime mortar, do not use cement.

throat

rubble on vent lines usually built on cant, 6" to 7" thick (150 to 175 mm) do not use strong cement mortar for rebuilding cants: stone on end.

inside elevation

Figure 55

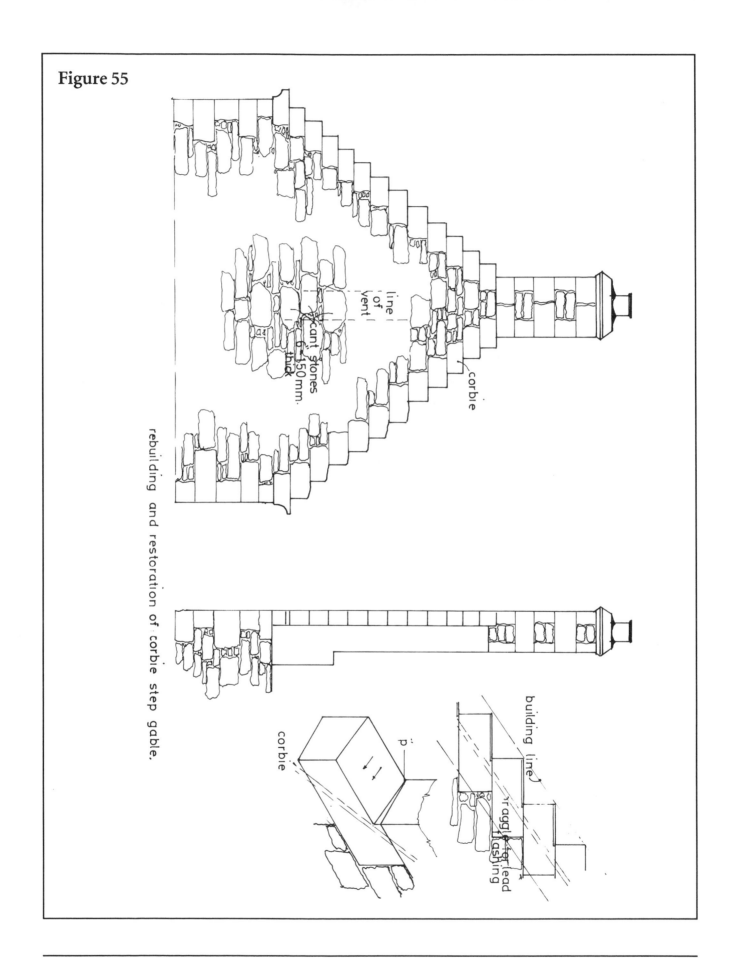

line of vent

scant. stones 6 - 150 mm. thick

corbie

rebuilding and restoration of corbie step gable.

building line

corbie

p

raggle for lead flashing

Figure 56

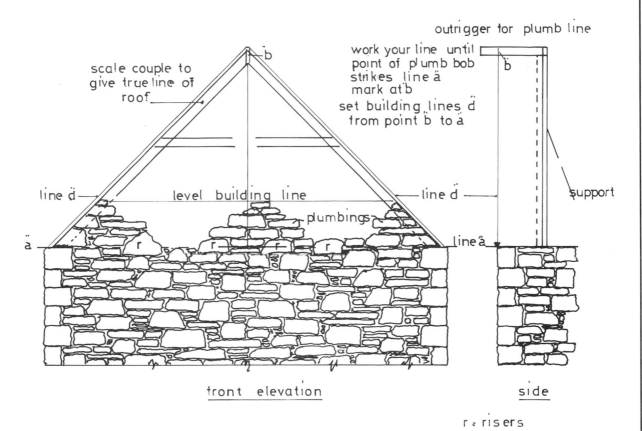

scale couple to
give true line of
roof

line ä

ä

level building line

plumbings

r   r   r   r

outrigger for plumb line

work your line until
point of plumb bob
strikes line ä
mark at b
set building lines d
from point b to a

line ä

support

side

r = risers

front elevation

setting off for building random
rubble gable.

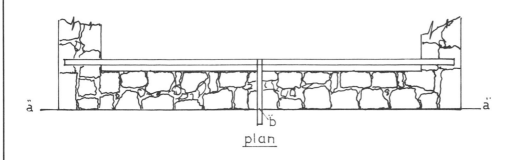

ä   ä

b

plan

Gothic Arches at Iona Abbey

I worked for a number of years doing restoration and rebuilding work on the historic and ancient Abbey of Iona, off the Island of Mull, Scotland. This is, ecclesiastically speaking, perhaps the most famous spot in Scotland.

The coursed random rubble on the Abbey and its adjoining buildings differed in style from anything I had undertaken before. It was a hard red granite, with a black slate infill; the sand used for the mortar mix and pointing was from the white sands of the seashore. All the corner stones for windows, doors, arches, etc., were of dressed sandstone.

It was during the latter part of my restoration work that I accepted the position of master-mason, building the Cloisters. Composed of four arcades, the Cloisters had a total of fifty-seven depressed Gothic arches. The restoration included cutting and building two large semi-circular corner arches. The other two corner arches were in good shape (Figure 57).

The layout had to follow the foundation line of the existing Cloisters. A few stones showed above the ground level. If you find yourself working on an old church such as this one, you will discover that nothing is at right angles or square, and no two sizes are the same. I spent a few hours going over the drawings, checking stone sizes against the stone available for each arcade, as on each arcade the span of the arch differed. With the layout drawing, I checked the sizes and levels against the existing structure. Then I worked out the procedure to be used in fitting it all together and got a picture in my mind of what the finished Cloisters should look like.

Next, I had to set all this out on site, using pegs and string. The leveling instrument I used for the entire job was a 10-foot board with a 4-foot level on top. Each arcade depended on strict accuracy. You may think, "What's a 32nd of an inch?" But thirty-two 32nds equal 1 inch, and you could be well on your way to disaster.

My procedure was to address the two large semi-circular corner arches first. I had to begin with these, as the butt of the arch formed the starting point for two arcades. I worked out my own templates, using the method described on page 68. With zinc moulds ready, I proceeded to dress or carve the stones to the required shape. The stone-dressing process must be done entirely by hand, with no saws or other machines to help out. Like all natural materials, stone has a grain and a bed and must be built the same way as it lay in the quarry. You work the stone so the pressure or compression will be at right angles to the face.

Having dressed all the stone required for this one arch — twenty-six stones — I first built the butt, or abutment wall, then fitted in the timber center required. Then I began to set the voussoirs over the timber center. To me, finding everything to size and plumb is the most satisfying part of building.

The next corner arch offered a little different challenge, as much of the original arch was standing to the halfway mark of the circle. This meant working the new into the old; again, I built a new abutment and center frame. I then completed this arch, other than the large stone gutter, which I had to cut and build after the roof was complete. The four corner arches were now fully restored; all that remained to address were the fifty-seven Gothic arches.

Having cleared and cleaned down to the top of the existing founds of the original cloisters, I started by building on top of the existing founds in random rubble. These walls were to carry the heavy arcade of arches. I carried through this

The west arcade of the Cloisters at Iona Abbey showing two original arches. The restoration was completed in 1959.

The north arcade of the Cloisters under construction, showing positions of sills, stools, columns, caps, and springers.

The north arcade under construction, showing voussoirs, spandril rubble, and the wall head course.

building, leveling and lining until I had completed the wall on the four arcades. It was now set for the delicate and time-consuming work to come.

I decided I would complete one arcade at a time; this strategy would allow the carpenter to carry on with the roof works while I continued with the next arcade. I began by setting the stone sills along the first arcade, bedding at the ends only where the load of the arch would be applied. Otherwise, I used hollow bedding, the usual method of setting sills. Jointing for this work had to be accurate, as the stools for the columns were to be set at this point. I leveled and checked at every point.

Two sample arches were built to determine how they balanced out, and to see whether the voussoirs on the arches looked too heavy for the slim structure. I found that they did look too heavy, so I reduced the height. This helped to bring down their weight, making them easier to

manhandle into position, as there was no lifting gear on this site. During the building of the sample arch, I learned you cannot allow any margin for error. At one point, I went off my line only the thickness of a sheet of paper, and it threw my key stone in twist to $3/4$ inch.

After the sample arches had been rectified and taken down, I was ready for the base stones, or *stools*, of the arcade (Figure 57A). The stools carry the two columns, with the position of the columns worked out on the stools. Small holes were cut on the center mark for the columns on the stools. Each hole was for a small piece of hard stone or gravel to act as a dowel connecting into the column. I felt this would give the column extra strength and stability.

After setting all the base stones or stools, we made a dry run to position the thirty columns. This was to determine that the columns fitted into position and were accurately sized. This had to be done on a windless day. Once the dry run was complete, the columns were taken down and laid aside for resetting. Before setting each column I cut a small hole on the bed joint to fit over the dowel bedded in the stool. On the top bed of the column I cut a small hole for a dowel to fit in. Once the column was in position, the cap would sit on this dowel (Figure 58). All was set on a lime-based mortar, with center buttons to give accurate bedding.

The columns themselves were difficult to plumb, as they were cut with an *entasis*. Each column had a bulge or convexity of an inch on the center of the column. This outward bulge creates an optical illusion, making the columns look plumb and pleasing to the eye. If the columns were cut straight, they would look hollow.

All the plumbings were carried out with the old-fashioned 5-foot long yellow pine plumb rule with a swinging plumb bob, the most accurate of all plumb levels.

Setting the columns on the lime was very time consuming, as they had to be accurate in level and position. The cap must fit perfectly on top; there is no way to adjust without taking the lot down. To line them up accurately, I had marked a center line on the tops of the columns, then stretched my string line from end butt to end butt, the center line being marked on the butt with the height. Therefore, it was only a matter of working the center line on the column to this stretched line. I completed this line of thirty columns, with no visual means of support. That is why I inserted the piece of gravel as a dowel in each column, to give the column something to grip.

Before setting the caps, I fitted my dowel into the holes already cut on top of the columns, bedding them in and forming the beds with buttons to carry the caps. I also cut holes in the cap bases to receive the dowels projecting above the columns. My lines were set and squared. I lifted the cap stone above the column, then lowered it into position on the prepared bed of the column, allowing the gravel dowel to go into the hole cut to receive it. Setting this stone must be accurate the first time. You cannot adjust by any movement or you will break adhesion between the column and the base stone, causing it to twist. This would mean taking the whole thing down and rebuilding.

Having set the fifteen caps and the two at the butt ends, the work was in the most vulnerable position. It was top-heavy, and a gust of wind or

Cloisters at Iona Abbey.

Cloisters at Iona Abbey.

A restored semi-circular arch in the north-east corner of the Cloisters, Iona Abbey.

a careless touch could have the same effect as tapping a line of standing dominoes. When one goes, they all go. To solve this problem, I laid long boards along the top of the caps, wedged them at both ends, and placed weights on top to prevent any movement.

With the columns and caps in position, the next stones to set were the springers. These are the stones from which the arch springs. I lined through and marked the position on each cap for the springer. For greater accuracy I left the line in position on the outside face. Again, this demanded a direct single lift into the position already marked, as you cannot adjust once it is set in position. The board on top of the caps was removed and laid aside for placing on top of the springers.

After setting the springers, the next phase was to make a timber center frame to hold up the voussoirs during the construction of the arches. I decided to build three arches per day, so a three-centered arch frame was made and placed in position.

I was now ready to set the voussoirs. Regard-ing the board on top of the springers: I kept cutting and adjusting it to allow three arches to be built. I set the lines and started setting the voussoirs on a lime bed with three buttons. Before lifting into position, I cut a joggle down each bed joint (Figure 59). This was for grouting and to tie the stones together once they were in position. Then the key stone was set. At this point I knew whether I had kept to lines and sizes. I carried on doing three arches per day until I had completed the first arcade. Then I filled the spandril of the arches in random rubble up to the bed line of the wall head course, checking lines and sizes constantly.

The other three arcades were followed through in the same procedure, except in the last arcade. Pieces of the original cloisters were found during excavations in the 1920s. These were put together at that time to show how the originals had looked. I gathered enough of these pieces to form two arches. These were built, forming two arches within the new arcade, and mixing the old with the new.

A restored semi-circular arch in the north-west corner of the Cloisters, Iona Abbey.

Figure 57

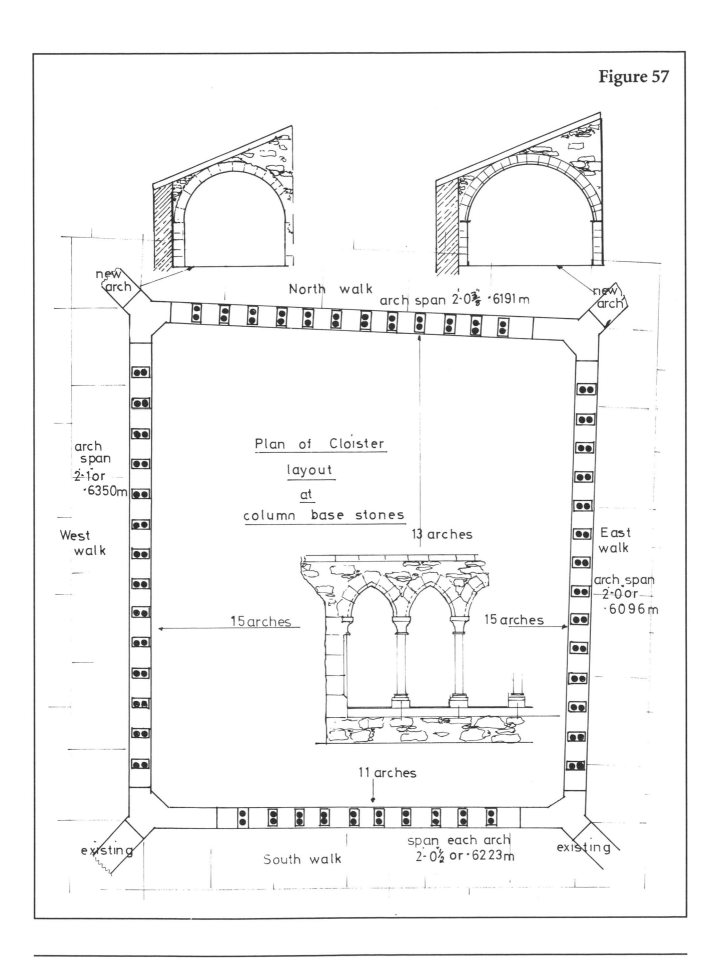

new arch

North walk   arch span 2·0⅜" ·6191 m

new arch

arch
span
2·1 or
·6350m

Plan of Cloister

layout

at

column base stones

13 arches

West
walk

East
walk

arch span
2·0 or
·6096 m

15 arches

15 arches

11 arches

existing

existing

span each arch
2·0½" or ·6223m

South walk

Figure 57-A

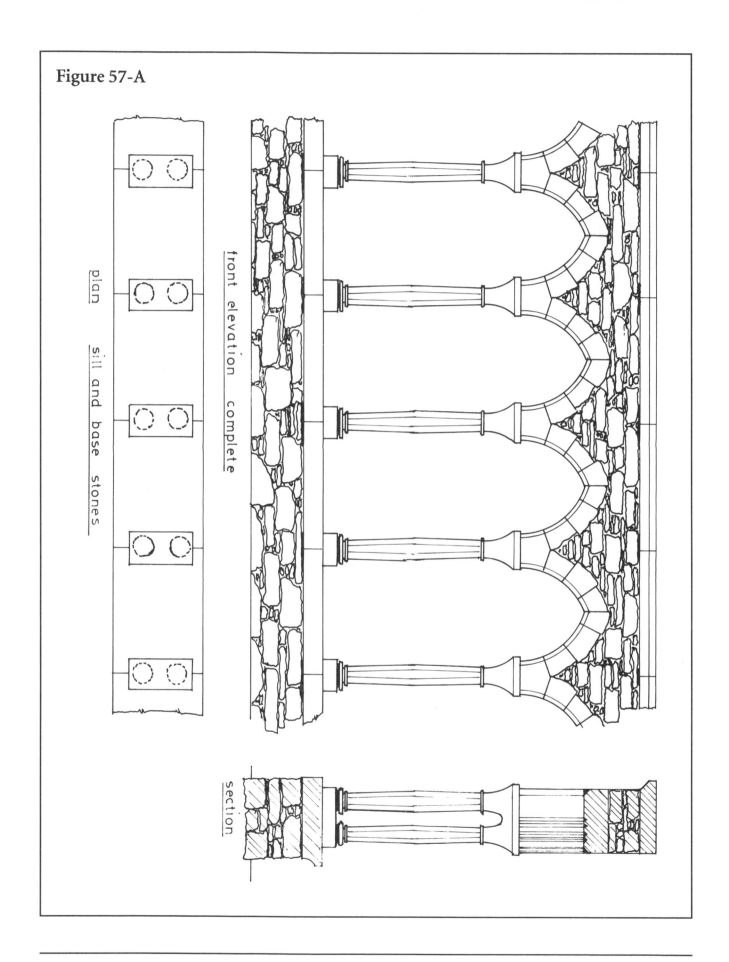

plan

sill and base stones

front elevation complete

section

Figure 58

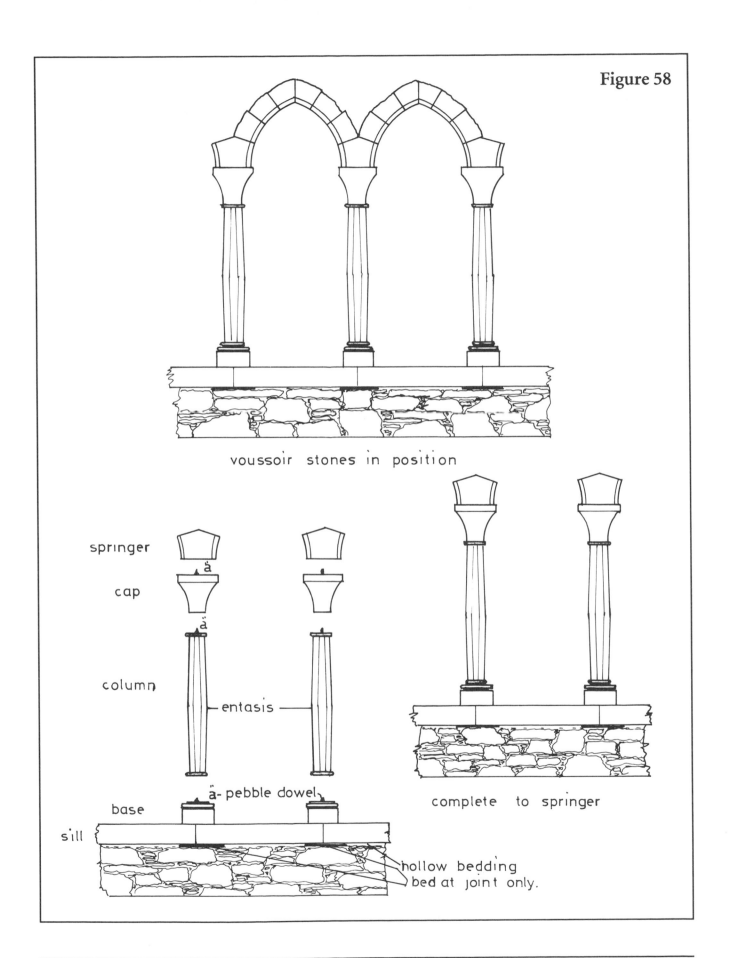

voussoir stones in position

springer

cap

column

entasis

base

sill

ä- pebble dowel

hollow bedding
bed at joint only.

complete to springer

Figure 59

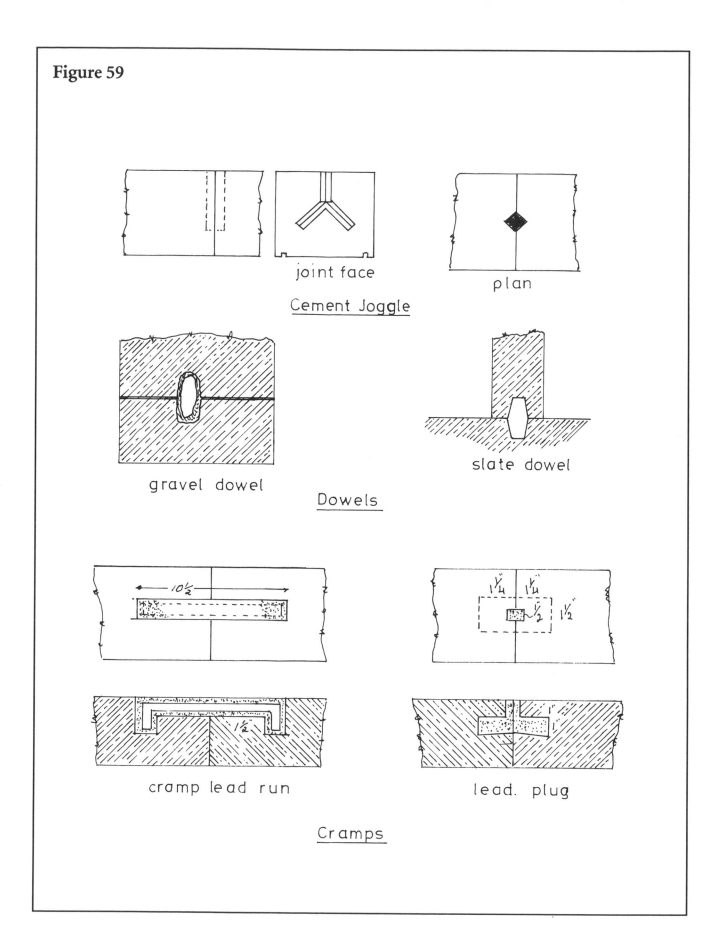

joint face

plan

Cement Joggle

gravel dowel

slate dowel

Dowels

cramp lead run

lead plug

Cramps

Stones Used in Masonry

For buildings of all kinds there is a great variety of stones from which to choose. Stones are of varying colors, structure, workability, and durability. All these qualities play important parts in the value of the stone as a building material. Their choice requires careful judgment and a fair knowledge of the constituents of the stone. The stones used by the mason for general construction purposes may be found in two geological classes: *igneous* and *aqueous.*

Igneous Rocks

Igneous rocks are of volcanic origin, having been formed under conditions of great heat. They vary considerably as to their chemical and physical composition. Igneous rocks include the granite types.

Granite is composed of quartz, feldspar, and mica. Quartz is a hard, glassy, flinty substance and is practically indestructible. Feldspar is crystalline and lustrous and comes in many colors including gray, pink, red-brown, and black. Mica is a transparent, silvery, scaly material, which lies in flakes parallel to the quarry bed. It is the weakest part of the stone's composition and should not amount to more than 5 percent of the whole.

Good granite contains about 50 percent quartz, a slightly lower percentage of feldspar, and the remainder mica. Granite is used for all purposes for which great strength is required or great weight is to be supported. It will stand enormous wear and hard use. Granite is used for engineering works, bridges, piers, foundations, and first floor facings of strong and high buildings.

Aqueous Rocks

Aqueous rocks are sedimentary rocks that have been formed or deposited in water. They are stratified or granular in structure. They are composed of particles detached from pre-existing rocks by the action of air, frost, and rain, or by force of contact with a moving body, such as a glacier. The particles are carried by the wind or washed by water into hollows, inland lakes, or seas. The water passes over them and deposits the grains. Of these rocks, those used for building purposes may be divided into three classes — grits, fine-grained sandstones, and limestones.

Grits are coarse-grained sandstones of great strength. Their roughness gives them their name. They are used for heavy engineering work, where resistance to wear and crushing is essential.

Sandstones are composed of grains of silica (flint or quartz) cemented together by silicic acid, carbonate of lime, carbonate of magnesia, or alumina. They also contain oxide of iron in varying proportions, and owe their different colors to the iron. A good sandstone should have an even, crystalline texture; it should be homogeneous with the strata not easily discernible, except in color. It should be clean, sharp, and bright when cut; a dull, earthy, or amorphous appearance denotes a tendency to decay.

Limestones consist of crystallized grains of carbonate of lime cemented together by the same material or a mixture of silica or alumina. They are of chemical or organic origin. Those specimens of chemical origin are formed by the action of carbonic acid on carbonate of lime, which is extracted from pre-existing rocks by water passing through or around them.

Limestones of organic origin are the result of minute organisms that extracted the carbonate of lime from the water in which they lived, making (with the lime) a shell-like covering or protection. When the organism died, the shell remains became cemented together, forming a mass of carbonate of lime.

Granular limestones are the class most used for buildings. They are called "oolites" because they are obtained from oolite or eggstone formation.

A good limestone has a dense, homogeneous structure and composition, with fine, small grains of uniform size and a crystalline texture.

Marbles are of this class. Their peculiar crystalline structure is caused by heat and compression, sufficient to cause alteration in form but not in constitution.

Stone Tests

The most reliable test for stone is to examine an old building nearby that has been built of the same stone. The *arrisses* (edges where the surfaces meet at an angle) should be firm and fine, and the members of moulds sharp and clean. The lines of stratification should not be prominent. The faces must be hard and solid when struck with a chisel. A loose or spongy appearance would denote decomposition of the chemical constituents.

The following are some specific tests for stone.

Water test — A few stone chippings are placed in clean water and stirred about. If the water becomes muddy, the stone should be rejected.

Chemical test — Immerse a stone in a solution of 1 cup sulfuric acid, 1 cup hydrochloric acid, and 1 gallon of water for a few days. When taken out and dried, the grains should be sharp and firm. Loose sand would mean the stone could dissolve in a polluted city atmosphere. NOTE: These acids are *very* dangerous. Every precaution must be used in handling and disposal.

To detect presence of lime — If a few drops of acid are placed on a stone and the drops cause effervescence, carbonate of lime is indicated. Such a stone would not weather well.

Absorption — A sandstone should not absorb more than 10 percent of its weight in water; a limestone not more than 17 percent.

General Characteristics of Stone

The best building stones are those obtained from the oldest formations or from the greatest depths in quarries, having been subjected to enormous pressure from the earth above.

Weight of stone — When stones are heavy and compact, with strong grains, they are generally used for heavy classes of work — engineering and marine.

Fine-grained stone — Fine-grained stone is lighter and finer in texture; it is more suitable for carving and ornamental work.

Porosity — All stones are porous to some extent, but some are so porous that they will not do for buildings that are in an exposed position, even though the materials are of good quality.

Seasoning — Stones are better for being seasoned before fixing. By exposing a stone to the air, all the moisture or quarry sap is dried out. Stone is more easily worked when green or newly quarried, but it hardens as it dries out. All the sap should be got rid of before the stone is fixed. Stone dries more slowly when fixed, giving acids a better chance to act on it and frost a better chance to crack it.

Natural bed — The natural bed is the bed on which the grains have been deposited in the quarry. Stone is softer on the beds, as the particles are flatter and it requires less force to detach them. All stones in walls, and especially those of a laminated character, should be placed with the natural bed at right angles to the force or the weight above. By this, the beds are protected; being built in, the grain or edge is the only part exposed. If a stone were placed "bed out," the particles would flake off. In a laminated stone, the layers or laminae would fall off like the leaves of a book, with the destroying agencies eating behind and forcing the layers away from the body of the stone.

Cutting Stone with Hand Tools

The following are the labors recognized in the preparation of stonework (Figures 60 and 61):

Scabbling — Taking off the irregular angles of stone with the scabbling hammer. This is usually done at the quarry, whereby it is termed "quarry pitched" or "hammer faced."

Hammer dressing — Roughest description of work after scabbling. The tool used for this is generally the wallers hammer.

Self faced or quarry faced — The term applied to the quarry face of the stone, or the surface formed when the stone is detached from the mass in the quarry; it also refers to the surfaces formed when the stone is split in two.

Chisel draughted margins — To reduce an irregular surface to a plain surface, a rebate about an inch wide is worked at two opposite edges of the surface. The parallelism of these rebates is ensured by testing with winding strips and a straightedge. A similar rebate is worked on the two remaining edges, connecting those first made. A continuous margin or rebate is thus formed about the four edges of the stone, every portion of which lies in the same plane. If the stone is small, the irregular excrescence is then removed with the chisel to the level of the rebate. For large stone surfaces, subsidiary draughts are formed, traversing the stone between the rebates. In walls, the stones with hammer dressed or rusticated surfaces have chisel draughted margins sunk about the four edges to ensure the accuracy of the work.

Plain work — Plain work is divided, for purposes of evaluation, into half plain and plain work. The term "half plain" is used when the surface of the stone has been brought or cut near to the true surface. This labor is usually placed upon the bed and side joints of stones in walling.

"Plain work" is the term adopted for surfaces that have been brought accurately to the true surface of the stone and are usually used to form an exposed surface. For sandstone or hard limestone, plain work includes a tooled stroke and chisel draughted margins; for soft limestones, a combed or dragged surface; and for granite, a smooth tooled face.

Comb or dragged work — This is a labor employed to work off all irregularities on the exposed surfaces of soft stones. The drag or comb is the implement used and consists of a piece of steel with a number of teeth like those of a saw. This is drawn over the surface in all directions, after it has been roughly reduced to a plane with the saw or chisel, making it approximately smooth.

Boasted or droved work — This consists of making a number of parallel chisel marks across the surface of a stone by means of a chisel, called a boaster, which has an edge about 2 1/2 inches wide. In this labor, the chisel marks are not kept in continuous rows across the width of the stone.

Tooled work — This labor is a superior form of the above, care being taken to keep the chisel marks in continuous lines across the width of the stone. The object of this and the preceding is to increase the effect of large plane surfaces by adding a number of shadows and highlights.

Pointed and punched work — The bed and side joints of stones are often worked up to an approximately true surface by means of a pointed tool or punch. The contact surface of the pointed tool is a point; the contact surface of the punch is usually a small rectangle. This technique is often used to give a bold appearance to quoin and plinth stones. Where so used it usually has a chisel draughted margin about the perimeter.

Sunk work — This term is applied to the labor of making any surface below that originally formed, such as chamfers, wide grooves, sloping surfaces of sills, etc.

Moulded work — Mouldings of various profiles are worked upon stones for ornamental effect. Mouldings are worked by hand or machine. In the former case, the profile of the moulding is marked on the two ends of the stone to be treated by means of a point drawing about the edge of a zinc mould cut to the shape of the profile. A draught is then sunk in the two ends to the shape of the required profile. The superfluous stuff is then cut away with the chisel, the surface of the two draughts being tested for accuracy by means of straightedges.

Vermiculated work — Usually done on quoin stones to give a desired effect, a margin of about $3/4$ inch is marked on the edge of the stone, and in the surface enclosed by the margin a number of irregularly shaped sinkings are made. The latter have a margin of a constant width of about $3/8$ inch between them. The sinkings are made about $1/4$ inch deep. The sunk surface is punched with a pointed tool to give it a rough, pock-marked appearance.

Furrowed work — Used to emphasize quoins, this consists of sinking a draught about the four sides of the face of a stone, leaving the central portion projecting about $3/8$ inch, in which a number of vertical grooves about $3/8$ inch wide are sunk.

Tools

Chisels are divided into two classes. The hammer-headed chisels have the section of their striking end made smaller to lessen the amount of burr. Mallet-headed chisels have a broader striking end, to avoid injuring the mallet.

As shown in Figure 60-A, other tools for stone-cutting include:

1. Pitching tool, a hammer-headed chisel with a very blunt edge, used for reducing stone.
2. Punch or cluerer, used with the hammer for removing superfluous stone in rough dressing.
3. Point, mallet-headed, for furrowed and sparrow-peck work.
4. Mallet-headed drafting chisel.
5. Hammer-headed chisel.
6. Mallet-headed boaster or scablar, from $1 1/2$-inch size upward.
7. Tooth-tool, mallet-headed, with 1- to 2-inch replaceable teeth.
8. Mallet, used for striking mallet-headed chisels, made of hardwood.
9. Mash hammer, for hammer-headed chisels.
10. Wallers hammer or katchie, for roughly squaring stones in rubble work.
11. Square and set square.
12. Bevel, formed of two blades of metal, slotted and fastened with a thumbscrew.

You will need a ruler for all measurements, along with a small brush for cleaning away stone chips and dust.

Stone-Cutting

The following is a brief description of the process of cutting a stone to shape by hand. Nowadays, machines are used to cut stones to all shapes and sizes, doing away with the work of the cutting mason. I use the hand-cutting method, especially on restoration work, in which you use individual stones of different shapes and sizes as replacements. The process described is the *freestone method* for the conversion of a rough block of stone to a shape that will fit into the building as shown by your drawing.

Select one surface of the stone as a surface of operation; just as in working a piece of wood, you would select one face and work from that face.

The surface will differ according to the finished shape of the stone. In some instances, you would select the face of the stone; in others, the bed of the stone.

When working the stone, the surface of operation should conform to the definition of a plane; it should coincide in every direction with a straight line, so that any surface can be squared from it.

Once you have chosen your rough stone, determine the size of your finished stone and the surface on which you will begin work. Mark a guideline on this surface, then use a hammer and a pitching tool to cut the stone to this line.

1. Cut corners "a" and "b" (Figure 62) with a mallet and drafting chisel.

2. Cut a marginal draft, say about 1 inch wide, working between "a" and "b". Test this for accuracy using your straightedge.

3. Point "c" is selected. It should be approximately square with the rough surface of the block.

4. Now cut into point "c" and work another margin between "b" and "c", as described in number 2 above.

5. You must now obtain the exact position of point "d". Place a straightedge on the draft "a" and "b" and hold another straightedge on the opposite surface of the block, so that the edge coincides with the position of corner "c".

6. These straightedges are now sighted through, and while keeping the edge correct with corner "c", the other edge of the straightedge, you raise or lower until the edges of both are parallel. This means your stone is *out of twist*, a process called *boning*. The process of sighting through is called "boning," a method of making sure the surface will be true and not twisted. When the surface is flat and true, it is termed "out of twist."

7. Once the position of "d" is determined, a straight draft or margin is now worked between "c" and "d". Be sure to test with your straightedge constantly as you work, until your draft or margin is complete.

8. Work the draft or margin between "d" and "a", which will complete the four marginal drafts.

9. To remove the superfluous stone enclosed by the margins, a series of furrows is worked along the surface with a hammer and a punch, parallel to the draft "a" and "b". Cut these furrows close to the required surface.

10. The surface is then toothed, using your tooth tool and mallet, in parallel drafts worked in the same direction as the furrows. Again test using your straightedge.

11. Once number 10 is complete, work the surface with your mallet and a boaster tool. This should also be worked parallel to the margin draft "a" and "b", with each draft being tested and reworked until it is correct before the next draft is boasted. Each draft is a guide for the working of the next.

12. Your straightedge should then be applied diagonally across the surface. If your surface is round in one direction and hollow in the other, it is proof that the surface is twisted. You now have to repeat the complete process from item 1. The lesson in this is to take your time and work the stone accurately.

After you have completed working the surface of your stone, you apply your moulds to this finished surface, marking with a scriber and then with a pencil to define the line more clearly. You then work the various surfaces, square from your already finished surface. (Remember, cut your stone to fit your square, not the square to fit the stone.) Start with the marginal drafts, checking and squaring all the time.

To work a voussoir (or arch stone), you need face and section moulds as described in the section on arches (Figure 36 on page 76). Select a stone of suitable size and work one face as previously described. Make this the face of your stone and the surface you will square and work from, and to which you will apply the moulds.

Mark the face mould and work the top bed surface, the vertical joint, and the two normal joints all square from the face, to the correct outline of the face mould.

Next, work the soffit, which should be tested using a reverse template of the curve of the stone.

Figure 60

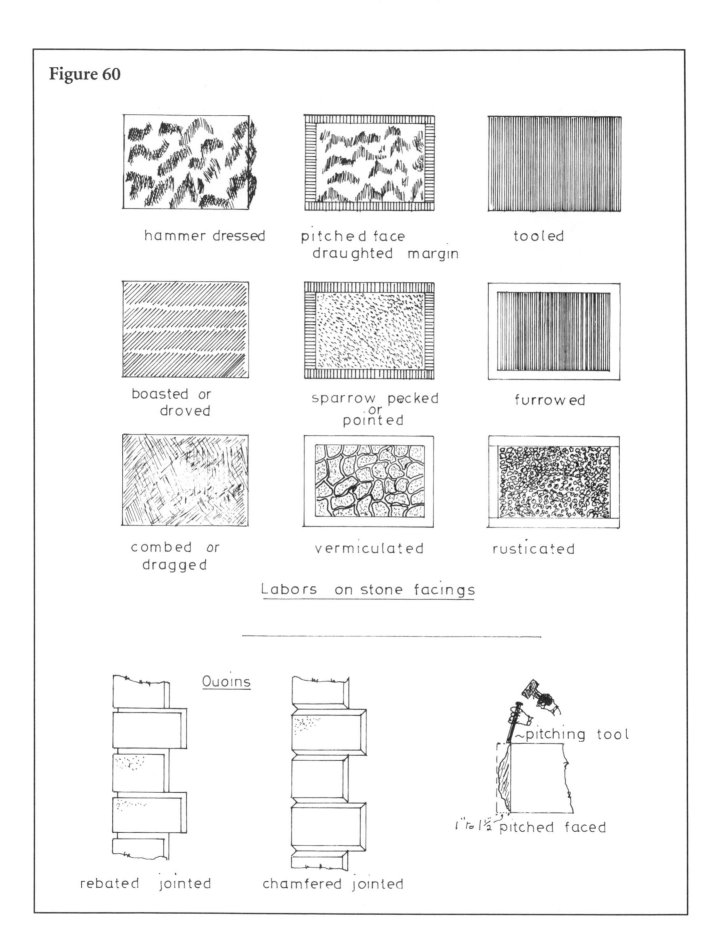

hammer dressed

pitched face draughted margin

tooled

boasted or droved

sparrow pecked or pointed

furrowed

combed or dragged

vermiculated

rusticated

Labors on stone facings

Quoins

rebated jointed

chamfered jointed

~pitching tool

1" to 1½" pitched faced

Figure 60-A

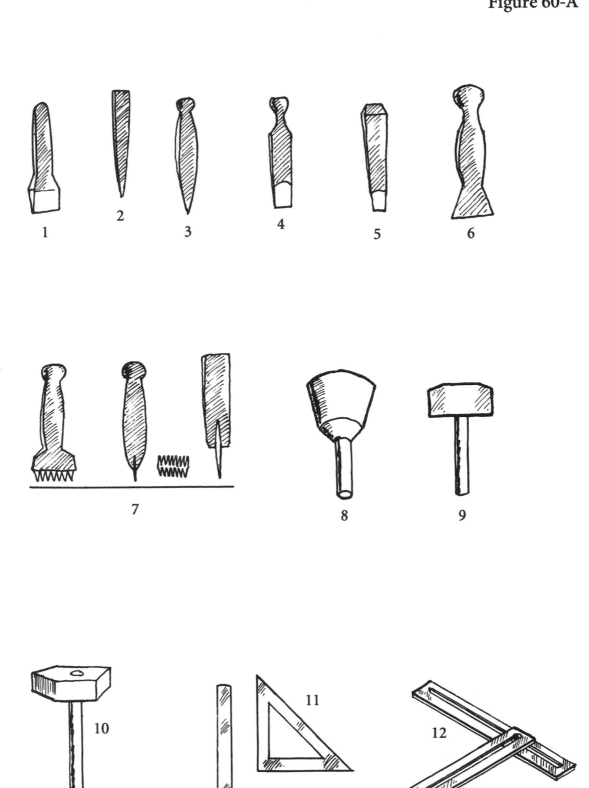

Figure 61

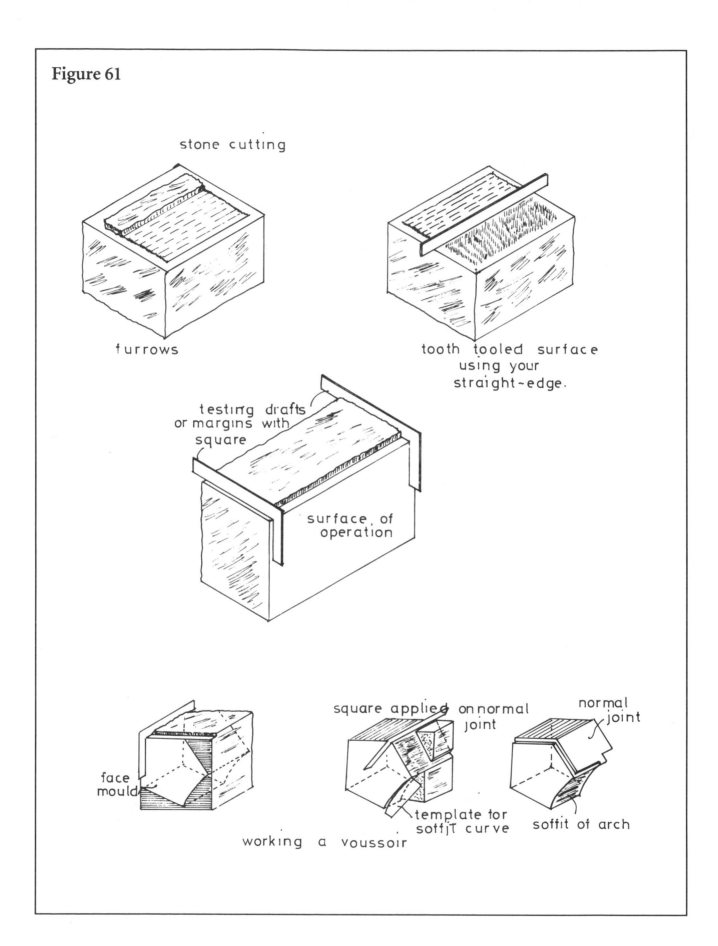

stone cutting

furrows

tooth tooled surface
using your
straight-edge.

testing drafts
or margins with
square

surface of
operation

face mould

square applied on normal joint

template for soffit curve

normal joint

soffit of arch

working a voussoir

Figure 62

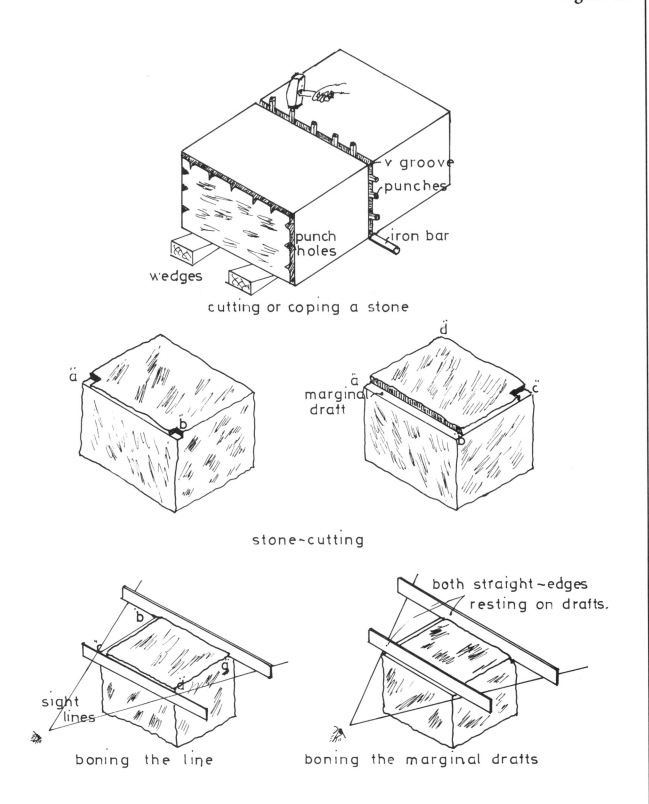

v groove

punches

punch holes

iron bar

wedges

cutting or coping a stone

ä

b

d

ä

marginal draft

c

d

stone-cutting

b

c

d

g

sight lines

boning the line

both straight-edges resting on drafts.

boning the marginal drafts

Stone Setting

For the following discussion, see Figures 63-73.

Stone setting means setting in position, per the building drawing, dressed cut stone, such as marble, sandstone, limestone, granite, or precast concrete panels. The method I use for stone setting, in which exact bedding and joints are required, I call the "button method." This is a traditional method of setting heavy stones.

First, we have to find the thickness of the beds and joints required, as the thickness of the buttons depends on the thickness. Prepare the buttons as shown in Figure 63. You will also need a number of small wood wedges (Figure 64). The buttons are to set your stone on; the wedges are for back wedging and for locking the stone in position.

The tools you will need are a mason's trowel; a 4-pound rawhide hammer, which is better than using a hammer and a block of wood, as you can hold the stone or your plumb rule with one hand and do the adjusting with the rawhide hammer in your other hand; a 2-pound club hammer; a small chisel for any extra tie holes; a small bar for adjusting stone; a bedding rod to use as your guide for clean jointing and bedding; an old saw with the teeth flattened to use for tightening up the joints by running the saw blade down the joint (do this only while the mortar joint is soft); and a string line for lining through.

For the purposes of our example, we will face the building with *ashlar*, 5 feet square and 5 inches thick. The foundations are in, so start by setting off the base course. This course must be built by setting the stone on a mortar bed, it must be accurate in measurements and level and plumb, as the rest of the setting will depend on this course.

Having set the base course, and completed the inner frame of the building (whether brick or concrete), you are ready for the next step. Remember to set the wall ties in position, as specified in your working drawing. Set the bedding rod in position (Figures 65 and 66), then spread the mortar bed. The mortar should finish to the top of the bedding rod. After spreading the bed, remove the bedding rod. The mortar bed should be about one inch from the stone face. Then set your buttons in 2 inches from the stone face and 4 inches from each end. The buttons must sit hard on the surface of the stone, not on the mortar bed. Now use the bedding rod as a screed or guiding strip for doing the upright joint. You do not need buttons on the joints. Lower the stone into position on top of the prepared mortar bed and buttons, adjusting for line and plumb using your rawhide hammer. The first stone is held in position with ties, and this stone will also act as your plumbings.

Move on to the next stone, remembering your ties. It is normal practice to fit joint dowels (Figure 67), so do this before you spread the mortar joints. Now carry out the same procedure as before. Having set the stone in position, use a small piece of brass bar as a gauge for the size of your joint. This will ensure accurate thickness of the joints. Once the stone is set in position and to the line, tap a small wet wedge into the top of the vertical joint. This wedge will hold the stone in position until you fit your ties.

Before using the wedges, soak them in water, allowing them to swell before placing them in position. Once in position, the wedge will dry and fall out, or you can break off the head of the wedge. On no account must wedges be used on the front face bed or joints or on moulded joints.

Carry on this procedure throughout, until you gain confidence in this simple method of setting stone. It is clean and fast, and gives greater accuracy in bedding and jointing, with no fear of settlement, no matter what the weight of the stone set above.

If mortar runs down the face of the stone, wash it off immediately using a large paint brush. Do not use a sponge or a rag, as this will only spread the stain. When the stone dries, the stain will be difficult to remove.

You should build up, point, and clean down. Do not attempt to point as you build. Your joints should be allowed to dry out. If you use the bedding rod correctly, all joints should be clean and all mortar back at least $^3/_4$ inch from the face of the stone.

Figure 68 shows the lifting apparatus for ashlar and precast panels not more than 5 inches thick. Drill a small hole $^1/_2$ inch deep by $^1/_2$ inch in diameter and 4 inches from the top bed of stone, dead center from both ends. Fit the button on the clamp plate into the hole you have jumpered, then tighten the clamp and proceed to lower the stone into position as described above, remembering your locking wedge once the stone is in position. Now release the clamp and remove. With ashlar or precast panels, you should have a minimum clearance of 1 inch between the back of the stone and the main structure of the building. This space is usually filled up with a mixture of 8 parts sand, 2 parts lime, and 1 part cement. This is a semi-dry mix, run and tamped into the space between the stone and the building once you have completed the stonework. Instead of using ordinary cement, use white cement, which prevents future staining of the stonework.

Squared ashlar, built 1860.

Squared ashlar, built 1860.

The author setting a marble star in the entrance to the Capitol Building, Washington, DC, 1960.

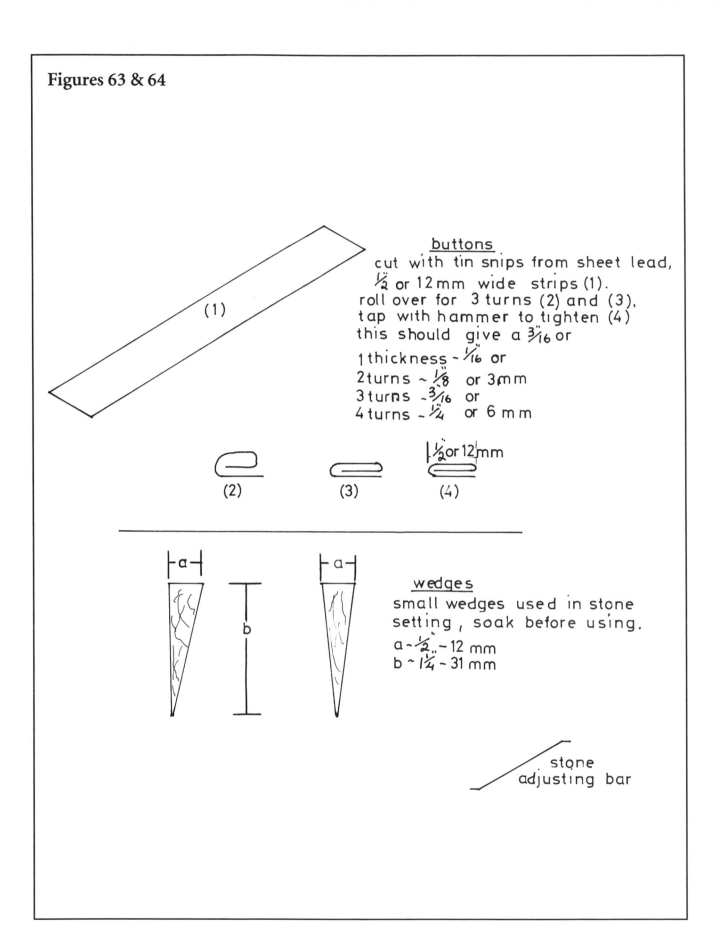

buttons
cut with tin snips from sheet lead,
½ or 12mm wide strips (1).
roll over for 3 turns (2) and (3),
tap with hammer to tighten (4)
this should give a ³⁄₁₆ or
1 thickness ~ ¹⁄₁₆" or
2 turns ~ ⅛" or 3mm
3 turns ~ ³⁄₁₆ or
4 turns ~ ¼" or 6 mm

(1)

(2)   (3)   (4)

½ or 12mm

wedges
small wedges used in stone
setting, soak before using.
a ~ ½" ~ 12 mm
b ~ 1¼" ~ 31 mm

a   a

b

stone
adjusting bar

Figure 65

STONE SETTING

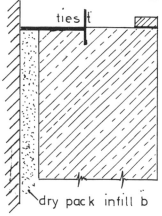

ties t

bedding rod in position.

method in using your bedding rod and setting button.

dry pack infill b

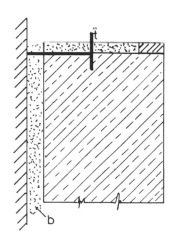

b

mortar bed, laid to level of bedding rod.

do not point as you build up, clean and point down.

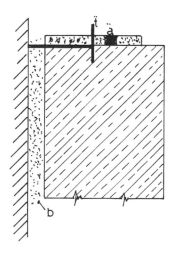

t a

b

a=button set on stone, not on mortar.

bedding rod removed, now ready for setting stone on top.

bedding rod can be used on vertical joints no buttons required.

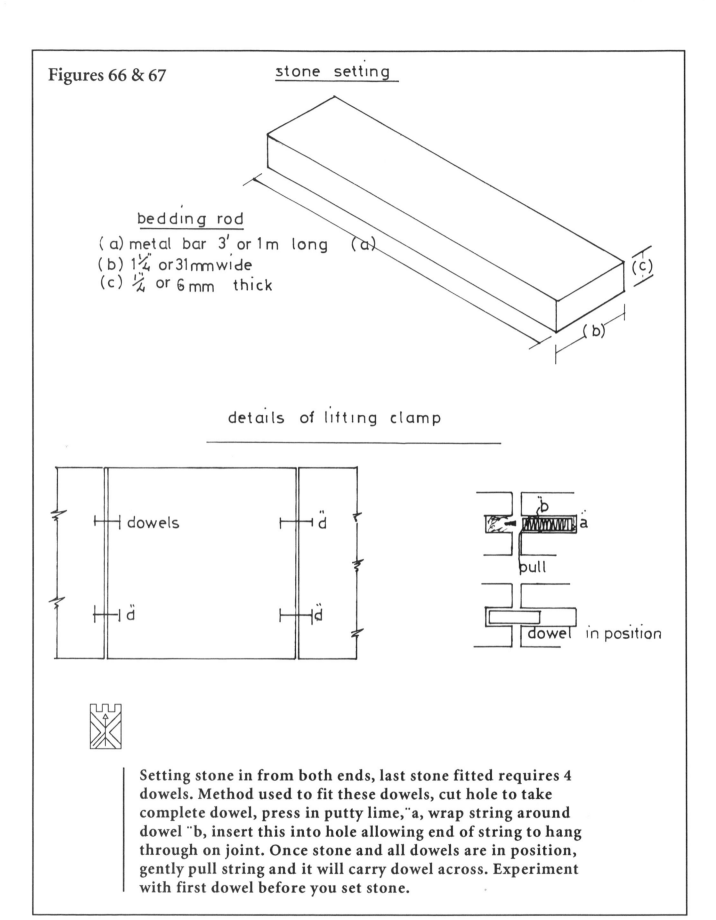

Figures 66 & 67

<u>stone setting</u>

<u>bedding rod</u>

(a) metal bar 3' or 1m long
(b) 1¼ or 31mm wide
(c) ¼ or 6mm thick

details of lifting clamp

dowels

pull

dowel in position

Setting stone in from both ends, last stone fitted requires 4 dowels. Method used to fit these dowels, cut hole to take complete dowel, press in putty lime, ˝a, wrap string around dowel ˝b, insert this into hole allowing end of string to hang through on joint. Once stone and all dowels are in position, gently pull string and it will carry dowel across. Experiment with first dowel before you set stone.

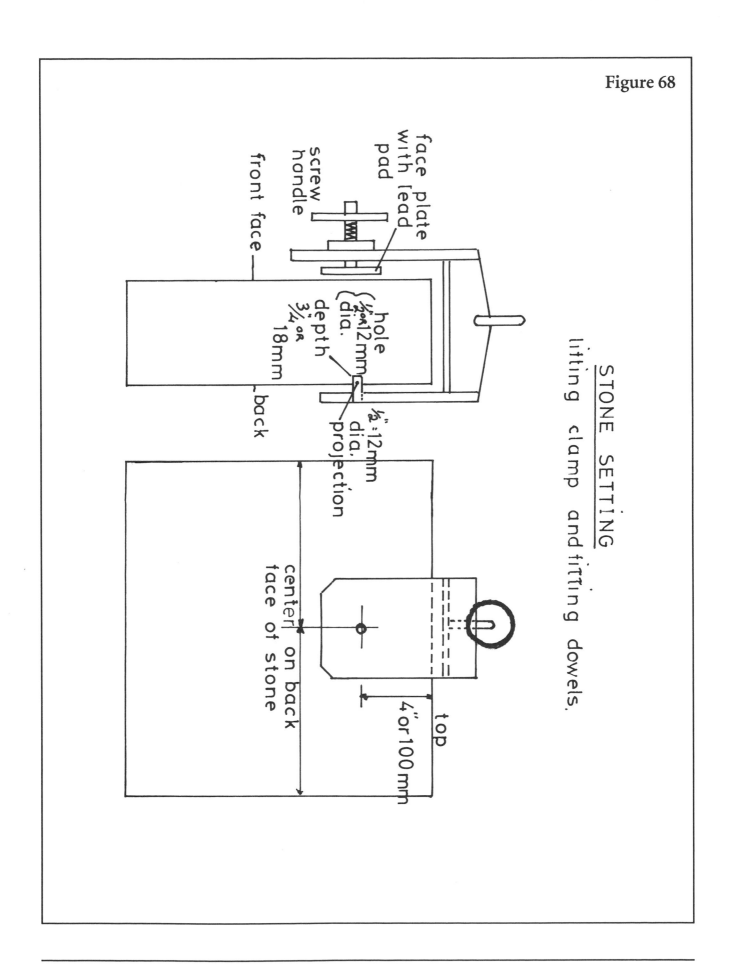

STONE SETTING

lifting clamp and fitting dowels.

face plate
with lead
pad

screw
handle

front face

hole
½" or 12mm
dia.

depth
¾" or
18mm

½" = 12mm
dia.
projection

back

center on back
face of stone

top

4" or 100mm

Figure 68

Figure 69

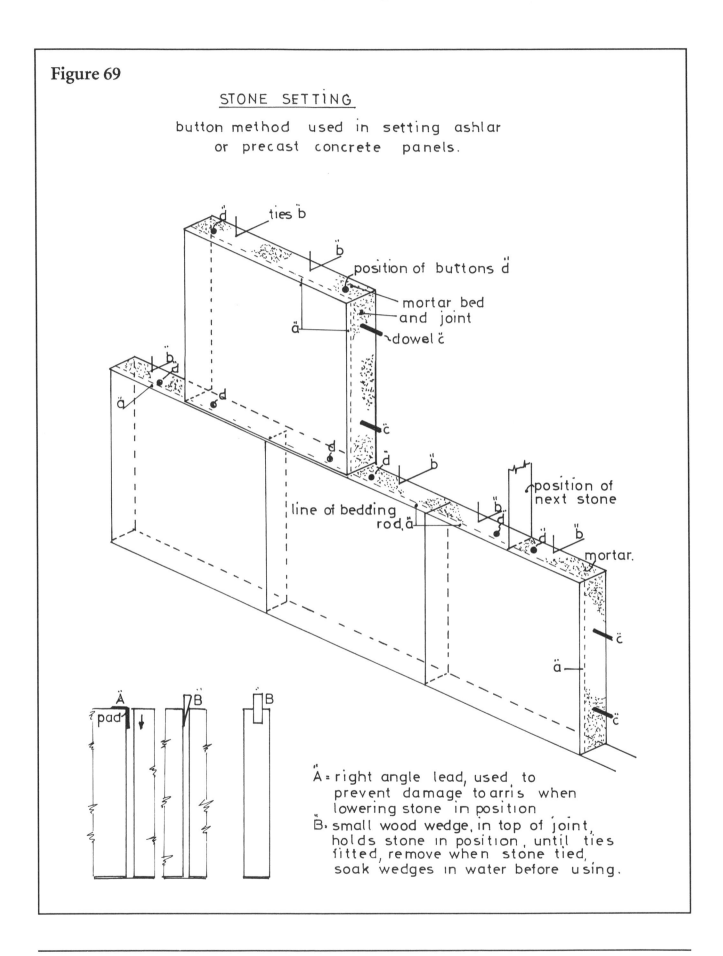

STONE SETTING

button method used in setting ashlar
or precast concrete panels.

ties b̈

b̈

position of buttons d̈

mortar bed
and joint

ä

dowel c̈

b̈ d̈

ä

d̈

c̈

d̈ b̈

position of
next stone

line of bedding
rod ä

b̈ d̈

mortar.

d̈ b̈

c̈

ä

c̈

Ä
pad

B̈

B

Ä= right angle lead, used to
prevent damage to arris when
lowering stone in position

B̈= small wood wedge, in top of joint,
holds stone in position, until ties
fitted, remove when stone tied,
soak wedges in water before using.

Figure 70

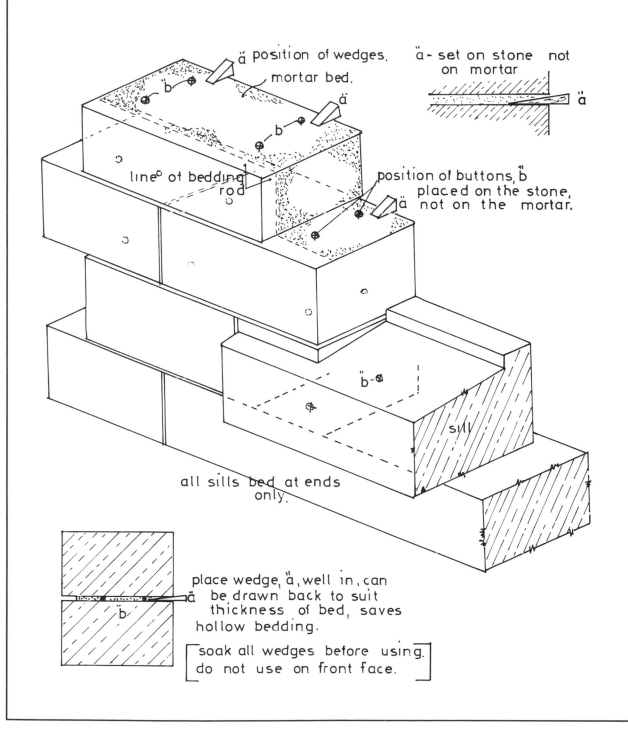

STONE SETTING

heavy masonry using buttons allows
height to be gained, without fear of
settlement.

ä position of wedges.
mortar bed.
b
ä

ä - set on stone not
on mortar

ä

line of bedding
rod
position of buttons, b
placed on the stone,
ä not on the mortar.

b-ä

sill

all sills bed at ends
only.

place wedge, ä, well in, can
be drawn back to suit
thickness of bed, saves
hollow bedding.

ä
b

[soak all wedges before using.
do not use on front face.]

Figure 71

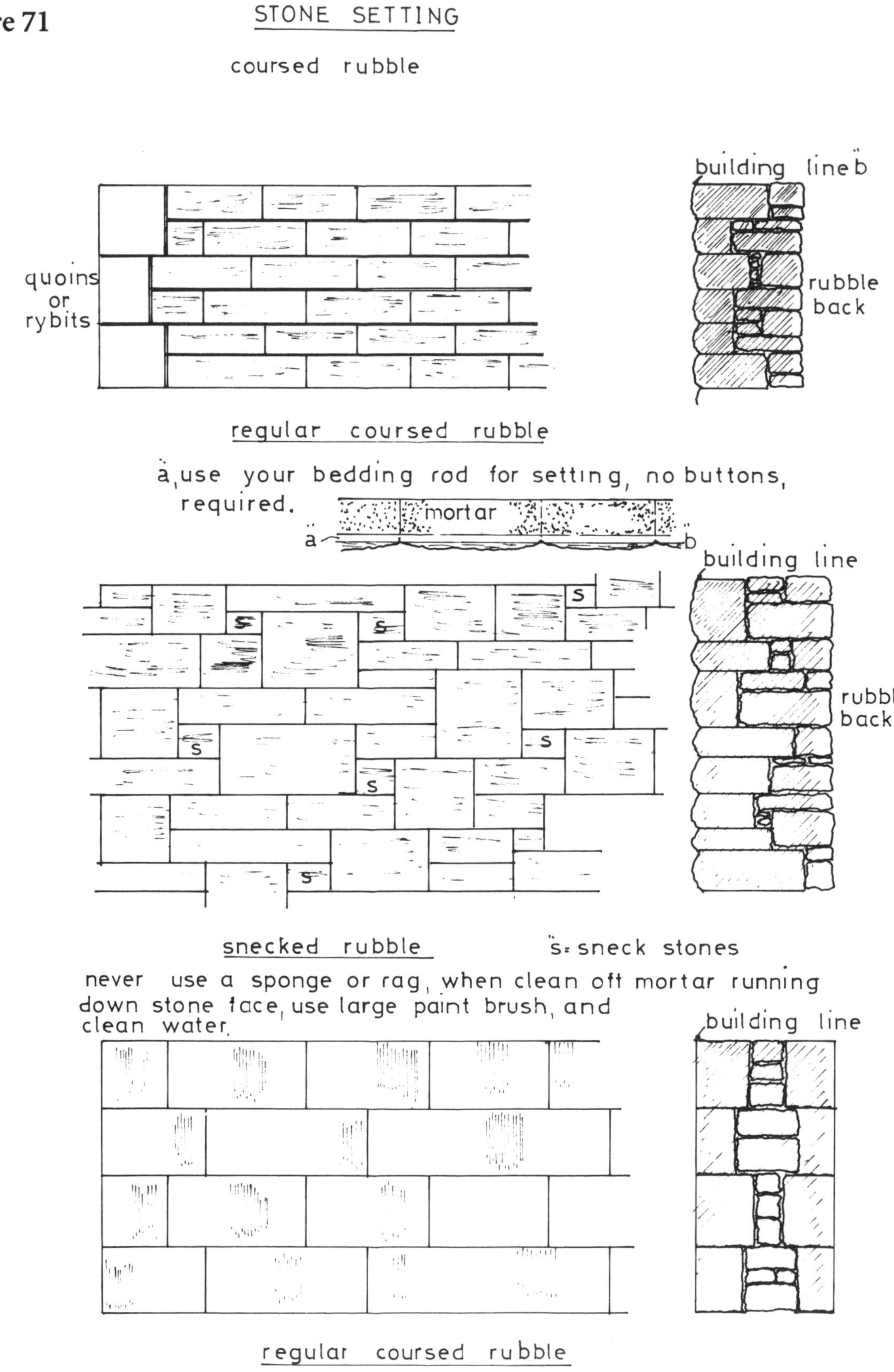

STONE SETTING

coursed rubble

building line b

quoins or rybits

rubble back

regular coursed rubble

a, use your bedding rod for setting, no buttons, required.

mortar

a   b

building line

rubble back

snecked rubble   s = sneck stones

never use a sponge or rag, when clean off mortar running down stone face, use large paint brush, and clean water.

building line

regular coursed rubble

Figure 72

stone setting

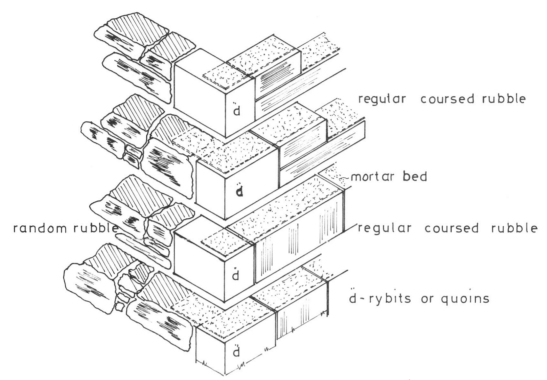

regular coursed rubble

mortar bed

random rubble

regular coursed rubble

ä-rybits or quoins

method of setting rybits or quoins
using bedding rod only.

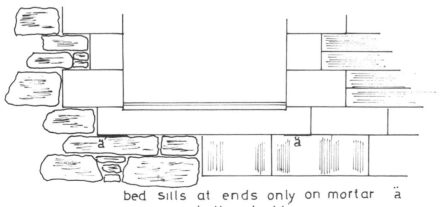

bed sills at ends only on mortar ä
hollow bedding

coursed rubble,-square cut stone, not exceeding 12″or 300mm in height.
ashlar-square cut stone, more than 12″or 300mm in height, joints not
exceeding ¼″ or 6mm thick.

Figure 73

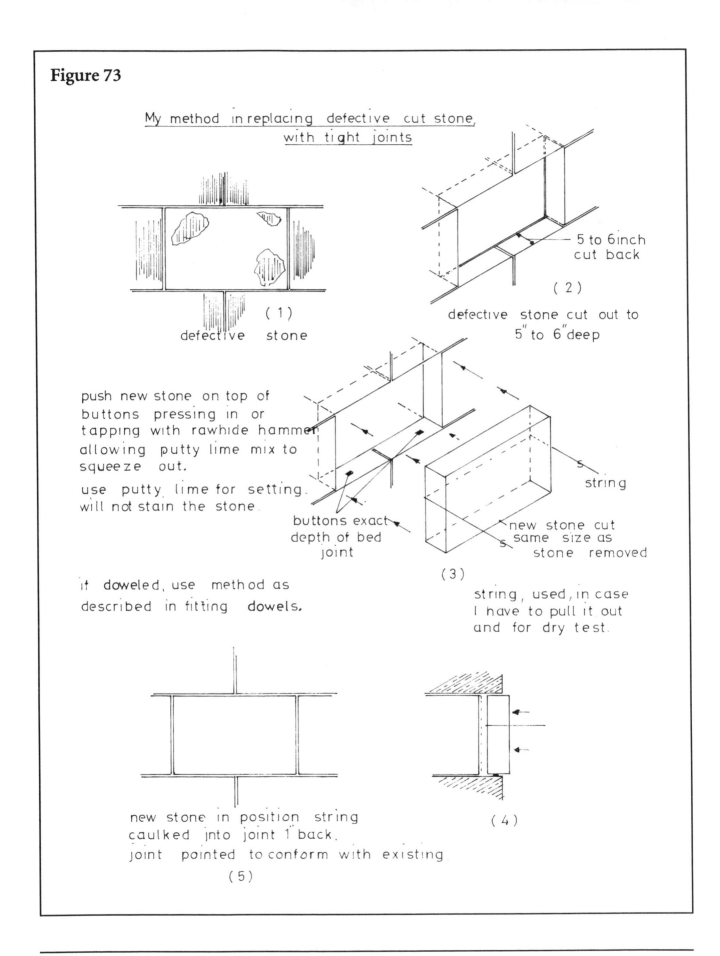

My method in replacing defective cut stone, with tight joints

(1) defective stone

— 5 to 6 inch cut back

(2)

defective stone cut out to 5" to 6" deep

push new stone on top of buttons pressing in or tapping with rawhide hammer allowing putty lime mix to squeeze out.

use putty lime for setting. will not stain the stone.

buttons exact depth of bed joint

string

new stone cut same size as stone removed

(3)

string, used, in case I have to pull it out and for dry test.

if doweled, use method as described in fitting dowels.

(4)

new stone in position string caulked into joint 1" back. joint pointed to conform with existing.

(5)

Preserving Stonework

The most effective method for preserving stonework, particularly in industrial areas where the atmosphere may be very polluted with soot and harmful gases, such as carbon dioxide, sulfuric acid, chlorine and ammonia compounds, is periodic washing and scrubbing of the stonework with cold or warm water. If the stone has been neglected over a period of time and is covered with a layer of soot, grime, and hard efflorescence from the stone, steam brushing may be used. Chemical applications should, as a rule, be avoided in this work.

Destruction of the stone face by frost may take place in porous stonework that is continually wet. A similar destruction will take place if, by use of an unsuitable mortar together with the presence of moisture in the stonework, efflorescence takes place continually on the surface of the stone. The removal of the cause of the dampness is the first and essential step.

The most effective surface preservation for dry stone work is ordinary oil paint, but since this destroys most of the surface appearance and character of the stone, it cannot be applied to important buildings. Boiled linseed oil is less objectionable in this respect, though it darkens the stone. Certain chemical preservatives are available that act with the constituents of the stone to form insoluble compounds in the surface of the stone. Only reliable and proven treatments should be considered. Where an important or historical building is concerned, careful investigation should precede its use.

Any treatment that closes the surface pores of the stone while the body of the stone is still wet may have disastrous results.

Exterior wall before repair.

Exterior wall from previous page after repair.

Interior wall showing repairs.

Pointing

In restoration and preservation of stone buildings, pointing is probably the most common operation carried out. Properly done, pointing is an important contribution to the maintenance and historical character of the building.

You will sometimes see a beautiful stone building "restored" with some colored cement mortar, raised out beyond the face of the stone. This method, in which the wall is built using small stones in the rubble, means more cement shows over the face of the building than does stone.

A much better approach is to use the traditional, time-tested materials and methods. These will help the building look as it did when originally constructed, which is obviously the purpose of restoring the building.

Modern materials now being used for pointing have not undergone the tests of time, and they are causing disastrous effects both in appearance and in hastening the deterioration of the stone. If the stone has been pointed with a strong cement mix, hairline cracking between the stone and the pointing will happen early on, allowing water penetration. After rain, the wall face will be dry but the joints will be damp, owing to the moisture trapped in the back of the pointing. You can test this by tapping the cement pointing with a metal rod. If moisture is trapped, the pointing will have a hollow sound, especially where the wall has been built with lime or clay. If you cut out a small section, you will see how easily it can be removed with little or no adhesion to the stonework.

Raised or V-joint pointing was a trend with stonework in the 1920s. It requires a strong cement mix for the depth required. Such strong cements are a fairly modern idea, dating to 1840 in England.

The use of colored cements and dyes on pointing stone should be banned. The colors are eventually absorbed into the stone, and within a very short period of time the building looks as though it has been given a fine coat of paint. Once these colors are absorbed into the stone, it is virtually impossible to remove them. Can you imagine what old buildings, such as cathedrals, castles, and mansions, would look like if colored cements were used?

If a building is constructed of hard slate or blue whinstone random rubble, with very tight horizontal joints and wide vertical joints, you can trace the joint problems back to the pointing method. If you cut out the face of the old pointing and scour out the joints with water, washing out the bedding material of the stone, you would find that repointing was done on the face only. No one seemed to believe in tamp pointing. With the joints so tight, you usually end up with a skin of strong cement pointing, which, within a short period of time, cracks and allows the weather to penetrate the wall, causing settlement.

Another pointing method I have found used on tight joints is cutting the stone to open the joint. This is only a face joint 1 inch deep and V-shaped. The cut creates joints about 2 inches wide. It certainly makes them easier to point, but it shows more cement pointing than stone and has no protective value or good adhesion.

Always tamp point joints, as this gives an equal depth in your finishing point. Otherwise, you end up pointing with some parts 1 inch deep and others 2 to 3 inches deep, which allows for unequal drying, with shrinkage problems on the deep joints. These are very difficult to finish off. To do so, you must cut back at least 3 inches into the existing joint, tamp point, then finish point.

The method I use on tight rubble joints, where it is impossible to tamp point and where there are a few slack or loose stones, is to take this

A house built in 1752. The existing pointing was cut to a depth of three inches, tamp pointed, and finished back pointed, to conform with the existing structure.

Antique flush point on top of joints already tamp pointed.

Flush back pointing, with antique pointing showing at top left-hand corner.

section out and rebuild it. I use a mix of 7 parts sharp sand, 1 part lime, and 1 part cement, no stronger. With this method, you make sure that the bedding and infill of the wall are solid. It makes it easier to point and gives the wall more stability. Repointing with only a face joint, with no proper bedding or infill and loose stonework, is asking for trouble in a very short time. Tamp pointing (Figures 74 and 75), unless carried out by skilled operators or under their supervision, is no guarantee that the pointing is solidly packed.

Another modern fault I have found is the use of a hose to force gallons of water over the stones and into the joints, washing out the bedding materials. I always use water sparingly. Large quantities of water are not necessary, and you should not soak the stonework prior to building. I only dampen the joints with a paint brush, prior to my finishing point. Deep tamping allows for any moisture in the lime-based mix to be absorbed into the stone.

I have never soaked a wall prior to pointing, and pointing I did twenty or thirty years ago is as good as the day it was done.

Tamp Pointing

When involved in cutting out joints and tamp pointing on defective stonework, always be aware of what is above and around you; use caution in all you do. Do not cut out more than you can tamp point in one day.

You should never cut completely around a large stone that is in position. If you remove all the defective pointing, the stone will fall out, allowing total collapse of a large area.

Cut and rake out all joints, removing growth and dirt, then brush down. *Do not scour joints with a hose.*

If there are small stones in and around joints, and they are loose or slack, remove them, clean out, and pack with your mix. Refit the stones in exactly the same position.

Your tamping mix should be as dry as you can handle. Push and tamp in your mix, consolidating it by using your tamper. Tamp point to within 1 1/2 inches of the stone face.

The tamping mix for lime-built walls is 7 parts sharp sand, 1 part lime, and 1 part cement. For tamping against clay or other materials (other than lime-built), do not use a strong cement mix. Your mix should be 8 parts sharp sand, 2 parts lime, and 1 part cement. When squeezed in your hand, the mix should hold up and not run. Your tamping will force out the moisture into the stone and the existing mortar mix. If the mixture is too wet, it will not hold up.

Do not use additives with mortar mixes, and do not remix an existing batch by adding water. Mix enough for your needs all at once.

Finish or Final Pointing

For finish or final pointing, the 1 1/2-inch to 2-inch deep joints should be filled up with a proper pointing mix, making sure it adheres to each side of the stone and the tamp pointing. Dampen joints only with a paint brush before starting to point. Remember to build up, point, and clean down, so pointing should be started at the top.

Surface pointing, or pointing any joint less than 1/2 inch deep, should be avoided, as it has no durability.

Your new pointing should conform in color to the old. The color of the sand determines the color of your finished pointing. Do not trowel or overwork your pointing.

Make sure the pointing is slightly recessed, not brought out over the face of the stone, which forms a thin skin over the edges.

My own method of finishing off the pointing requires a degree of skill and practice.

After pointing or filling up the joints, I let them dry out for one to two hours, depending on the weather. I then go over the pointing with a small piece of wood, giving it a rough finish. Then, using a fine spray of water, I use a paint brush to remove the film or skin of cement or lime, exposing the aggregate. This process also assists in cleaning and sealing the joints between the stone and the pointing and gives a good, weathered surface. Great care and patience are required for this method. Do not use a strong spray of water, and do not overwater, as this has a tendency to scour out the mortar and start it running down the wall. Do small areas at a time, starting from the top and working down.

In a second method, follow the same procedure as described above, but do not spray joints. Use a 2-inch wide paint brush and a bucket of clean water. Dip the brush into the water, remove it, shake it once to remove the excess water, then brush down over the surface of pointing, exposing the aggregate. Shake the brush to remove the material brushed off, dip it back into the water, then repeat the process.

A few other traditional methods are used in pointing stonework. Using the trowel to press the mortar in is called *shag pointing*, and it requires some skill. First, the joints are filled using the trowel, then when the mortar has dried out a little, it is rubbed over with a pad, usually cut from an inner tube or a piece of sacking. These give a good surface texture.

The other method is to point, and flush using the trowel, but instead of the pad, you draw the edge of the trowel over the surface, giving it a rough texture (with an old-fashioned look) when it dries.

To take away the bare effects of stone and mortar joints, it is common practice to mark some of the joints, either with a 1/2-inch half-round piece of metal, forming a keyed joint, or by marking a level line with the point of the trowel, then using the edge of the trowel to go over the line.

To give a rustic effect, you can avoid pointing, especially on garden walls and the like. When the wall is built, allow the mortar to dry out overnight. Scrape the joints in about 3/4 inch to 1 inch down the face of the stone using a small piece of wood. Then clean down and you can see the real effect of the stonework.

I have used another method on joints that are tight (to $1/2$ inch wide) on exposed stonework. Instead of adding water to my mix, I added boiled linseed oil. The joint was painted first with linseed oil before applying the mix. If a stone has been cleaned or refaced, painting with boiled linseed oil prevents pollution from penetrating the stone and allows it to breathe.

Finish Pointing Mixes

All pointing mixes must be measured accurately for uniform color, even up to the amount of water used. Hand mix on a board, using enough water to dampen, then work up the mix using the back of your shovel to bring out the moisture. The more water you add, the weaker the mix. Avoid machine mixing or mixing in a tub, as with these methods you tend to add more water to make the mixing easier.

Mix enough wet pointing mix to keep you going. Do not mix a large wet batch and leave it lying about all day, rewatering it every time you go to use it. A rule of thumb is to use the mix within a half hour of mixing it up. The mix should be 6 parts sharp sand, 1 part lime, and $3/4$ part cement. The color of the sand will determine the final color of the pointing.

Here is a recipe for an antique pointing mix. Take your sharp dry sand, put it through a fine sieve, laying to one side the fine sieved sand and to the other side the remaining gravel or grit. Then make up one of the following mixtures: 2 parts fine grit, 3 parts sieved sand, 1 part lime, and $3/4$ part cement; or 1 part $1/4$-inch gravel or grit, 5 parts sharp sand, 1 part lime, and $3/4$ part cement.

The above method is time-consuming in its preparation. All limes are hydrated; do not use additives, dyes, or any other methods of coloring the mixture.

Figure 74

TAMP POINTING

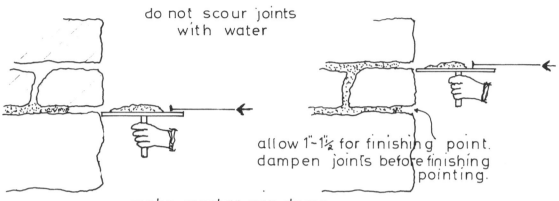

do not scour joints
with water

allow 1"-1½" for finishing point.
dampen joints before finishing
pointing.

make mortar mix damp
enough to handle.
push mortar into joint using tamper.
forcing the moisture in your lime
based mortar into the stone.

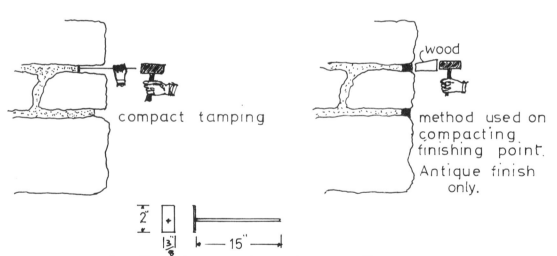

compact tamping

wood

method used on
compacting
finishing point.
Antique finish
only.

length of tamper depends on wall
thickness.

note
please read instructions carefully before attempting
to do tamp pointing.

Figure 75

<u>tamp pointing, pointing, crack repairs.</u>

 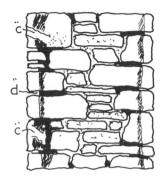

ä remove small infill stones, clean out opening, and rebuild in position as they came out

b̈ clean out and tamp point, doing small sections around each stone, do not remove pointing completely around stone, this will affect stability of stone and surrounding rubble. tamp pack this section before repeating the next section

c̈ leave small openings through your pointing for pouring in grout if required

d̈ allow 1"-1½(25-32mm) back from face of stone to allow for finishing point.

use only lime mortars

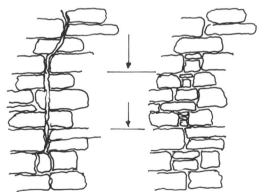 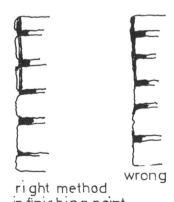

crack, work from top down, sections at a time, tying over, tamp pointing, and lime grouting.

right method in finishing point.

wrong

Mortar Mixes

The following mortar mixes are used for building with stone in traditional random rubble work as described in this book. Do not use strong cement-based mortars. All mortar used in building stone acts only as a binder. It is not the mortar that holds the wall together but the skill of the stonemason, especially in building random rubble.

The lime content of the mixture gives your mix flexibility and better adhesion to the stone.

General Purpose Mix

7 parts sharp sand
1 part lime
1 part cement
or
8 parts sharp sand
2 parts lime
1 part cement

Mix for Rubble Arches, Infill of Arches, Rubble Copes, Fireplaces, etc.

5 parts sharp sand
1 1/2 parts lime
1 part cement

Mix for Bedding Copes, Thin Sloping Copes, or Thin Cut Paving Slabs

3 parts sharp sand
1 part lime
1 part cement

Mix for Stone Setting

6 parts fine sharp sand
1 part lime
1 part cement
(Note: If building limestone or marble, use white cement.)

Mix for Grouting

3 parts fine sharp sand
1 part lime
1 part cement
(Thoroughly mix dry and add to your water until the required consistency for pouring.)

My own method is to mix the sand and lime together, bank it up and leave it, even for a week (Figure 76). Then, when ready to use, add the cement to the sand/lime mixture. You will find this sets up better on the stone and does not run.

How It Was Done in the Past

In the bygone days of building with stone, especially rubble work, and before the use of modern cements, graded and washed sand, and refined lime that comes in bags, what did stonemasons use as mortar in building castles, houses, and other stone structures?

The first part of the stonemason's apprenticeship was spent learning the ways of the laborer. The apprentice was taught the mixes and methods of mixing. There were no machine mixers; you hand-mixed on a hard surface, turning the mix over twice dry and twice wet. Very little water was added; the more water you add, the weaker the mix. You had to bring out the moisture with hard work.

I had only one experience with the old method, handed down through the generations. I remember going to a sand bank and hand shoveling out a load of sand — sharp and dead sand mixed through it. I think now that all the good is washed out of sand. My load of sand was taken to the yard and dumped. For every five tons of sand, two tons of hard shell lime was dumped on top. Lime is a substance produced by heating limestone to a high temperature, so the carbonic acid and moisture are driven off. The result is sometimes called *quicklime* or *caustic lime*. When you sweated and the fine lime dust settled on you during the hand-mixing process, you received bad burns. It was like a load of white clinkers, with unburnt stone through. The next process was to turn the sand and lime over twice dry and then bank it up, leaving this so the lime would slake down into the sand. The process took from three to six weeks. After a thorough inspection confirmed that the lime had slaked down, the whole lot was passed through a 1-inch to ³/₄-inch sieve, all by hand. After this was completed, the mixture was again banked up and the whole outside surface battered down with a shovel, which formed a protective skin.

This mix could lie as long as one wished. Whenever you were going to build, you cut down what was required. It was like cutting into thick butter. You dampened it and you were ready to start building. This was an ideal mix for stonework — it didn't run down the face and was used both in building and in pointing. The real proof and guarantee of this old method is seen in all the beautiful old buildings still standing.

A later development was Arden lime, which was an unrefined brown hydrated lime that came in bags. This you could mix with your sand, twice dry and twice wet. It was used for building and pointing. The mixing for pointing was hard work. You added enough water to dampen the mix, then battered it with the back of your shovel or danced on it until the moisture was brought out and it was damp enough to handle. Once you had the mixture in the joints, you finished off by tamping with a piece of wood and using a hammer to compact the joint. This mixture could also be mixed up in a large batch, left for about a week, and when you were ready for it, cut down and for every eight shovels you added one part of cement.

This was an ideal mix for building, especially for fireplaces and other general stonework.

The mixtures already described were time-consuming in their preparation, but labor was cheap and plentiful in those days, and there were no other methods.

In setting heavy dressed stonework, rybits or quoins, voussoirs, etc., where tight bedding was required, a putty lime was used. Putty lime was shell lime placed in a bin; water was added to it to cover the shell lime. The lime was allowed to slake down, which usually took about three weeks. After it was completely broken down, it was stirred into a slurry by adding more water. Then it was passed through a fine sieve into another bin and allowed to settle. After this had settled, it was ready to use.

Putty lime could be used for pointing, tight joints, and tight bedding stonework. Sieved fine sand would be added to the mix if the beds were thicker, with a button being used to prevent the stone from settling. Putty lime was also used for plastering.

Nowadays, the lime comes refined, doing away with all the preparation that was required in the old methods. When using the modern lime, especially on pointing mixes, which I mix by hand, I still allow the lime mix to cool for ten minutes before adding the cement. This seems to give a better feel to the mix.

All these methods I have described are from my own experience and from my own area in Scotland. Others may have used different methods in preparation, but basically the end result would be the same.

The modern mason will tell you these mixes with lime are not strong or hard, but from experience in cutting into old walls — 400 to 500 years old — I can tell you it is very hard work. It reminds one of cutting into a large, thick lead pipe, with no give. That is the *real* mortar and the way it should be with all stonework.

Figure 76

HAND MIXING MORTAR

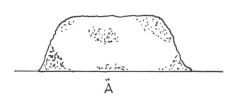

Ä

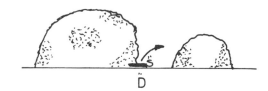

B̈

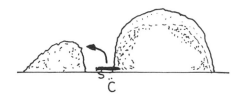

C̈

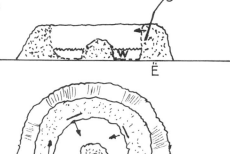

D̈

A ~ sand with top levelled off.

B ~ lime or cement on top use your shovel,s to lift tails.

C ~ 1st. turn over.⎫ turn over twice dry.
D ~ 2nd turn over.⎭

E ~ level out top, and hole for water, w.

F ~ use shovel, working around, pushing dry mix through water to centre, c

G ~ once F complete this is the batch ready for turning.

H ~ 1st turn over ⎫ turn over twice wet.
I ~ 2nd turn over ⎭

J ~ batch ready for use.

use a clean hard surface for working on.

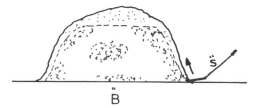

Ë

F̈

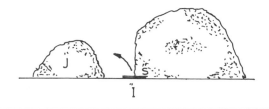

G̈

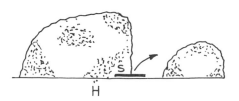

Ḧ

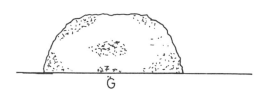

Ï

Time-Tested Materials

Lime

Lime is obtained by burning limestone. The characteristics of the lime depend upon the nature of the limestone. Thus "pure" limestone may yield almost pure lime or calcium oxide. Such a lime is obtained from chalk and the oolitic and mountain limestones. In the latter, however, there may be small proportions of magnesium carbonate.

Pure, rich, or fat limes are generally classed as non-hydraulic limes. They slake rapidly in water, increasing to twice their original bulk, and do not harden with time. They are chiefly used in plastering. They can now be obtained as a hydrated lime in a powdered form, which does not need to be slaked before use and can be stored for an indefinite period.

Slaking the lime was the old process in which one mixed sand with the freshly burnt lime. The slaking was done by adding water or leaving and allowing the wet weather to do the work, through which heat and steam are produced and the hydrated lime breaks down to a fine powder, mixing through the sand. It is very important that there be no unslaked particles throughout this mix.

Poor or lean limes slake slowly and do not increase much in volume. They are similar to fat limes in their other characteristics. In slaking poor or lean limes, the heat should be conserved by banking up the wetted lime with a layer of sand.

Boiled Linseed Oil

Boiled linseed oil is raw linseed oil so treated that it absorbs oxygen at a much increased rate, and thus becomes converted into a tough skin or solid in less time. The process is brought about by heating the oil to a high temperature and is increased by the addition, during the heating, of certain substances known as driers.

I have used boiled linseed oil as a pointing material on exposed joints, and for sealing stone from pollution. Mixed with mastic sand, it is an ideal sealant between stone and any other material. It is much used as a pointing material around doors and windows. In building stone during the industrial period and in building railway stations in cities, each stone was coated with boiled linseed oil before being set in position. This protected the stone from pollution penetration.

Water-Glass

Water-glass, or silicate of soda, is a concentrated and viscous solution of sodium or potassium silicate in water. It can be used as an adhesive, as a binder, and as a protective coating in waterproofing and preventing dusting of cement. It is an ideal sealer for stonework, because it allows the stone to breathe and is easily applied. The stone must be thoroughly cleaned either by steam or clean water, with no chemicals used. Water-glass is said to resist acid penetration.

Resources

Books

Arnold, Bob. *On Stone: A Builder's Notebook.* Origin Press, 1988.

Conservation of Historic Stone Buildings and Monuments. Washington, DC: National Academy Press, 1982.

Fine Homebuilding Staff, ed. *Building with Concrete, Brick and Stone.* Newton, CT: The Taunton Press, Inc., 1989.

Jerome, John. *Stone Work.* New York: Viking Press, 1989.

Kennedy, Stephen M. *Practical Stonemasonry Made Easy.* Blue Ridge Summit, PA: Tab Books, 1988.

London, Mark. *How to Care for Old and Historic Brick and Stone.* Washington, DC: Preservation Press, 1988.

Sanders, Scott R. *Stone Country.* Bloomington: Indiana University Press, 1985

Watson, Lewis, and Sharon Watson. *How to Build a Low-Cost House of Stone.* 5th ed. Stonehouse, 1974.

Wykoff, Gerald L. *The Techniques of Master Stone-Setting.* Adamas Publishing, 1986.

Government Publication

Grimmer, Anne E., compiler. *A Glossary of Historic Masonry Deterioration Problems and Preservation Treatments.* Washington, DC: U.S. Government Printing Office, 1984. (Write Superintendent of Documents, U.S. GPO, Washington, DC 20402.)

Articles

Holland, R. "Well-Wrought Walls." *Audubon,* May 1990.

McRaven, C. "Building a Stone Fireplace." *Country Journal.* November/December 1990.

Phair, M. "Steps of Stone." *Home Mechanix,* April 1990.

Vivian, J. "Seamless Perfection: Building Mortarless Stone Walls." *The Mother Earth News,* October/November 1991.

Wicks, H. "How to Build a Stone Wall." *Workbench,* July/August 1990.

Glossary

APEX STONE—The highest stone of a gable, cut to form the termination of two adjacent inclined surfaces. Also called SADDLE STONE.

ARRIS—The line or edge on which two surfaces forming an exterior angle meet each other.

ASHLAR — Stones that are carefully worked, usually over 12 inches in depth, and have joints not more than 1/4 inch thick.

BAND COURSE — A flat, horizontal stone, occasionally ornamented, dividing a wall surface.

BATTER — An upward and backward slope of the outer face of a wall.

BATTERED, or TO BATTER — To beat persistently or hard.

BATTLEMENT — A parapet with indentations, surrounding a wall; used for defense or decoration.

BEAM COMPASS — See TRAMMEL.

BED, MORTAR — The lower surface on which the stone rests and the upper surface that supports the stone above.

BED, OF STONE — The surface upon which the stone was originally deposited.

BEDDING ROD — Metal bar used as a screed for forming mortar bed on stone, as in stone setting.

BEE'S FEET — Slang for small stone used in random rubble.

BOND — The arrangement of stones so that the vertical joints of one course do not fall over the vertical joints of the courses above and below.

BONING — Making sure the surface being worked will be true and not twisted.

BUTTONS — Used on the beds for setting stone on.

CANT — Stone built with its natural bed exposed, on edge.

CASTELLATED — Having battlements like a castle.

CLUB SKEW — See SKEW CORBEL.

COPE — To split or cut a stone in a required direction.

COPING — The capping or covering of a wall to prevent water penetration through the top of the wall.

CORBEL — A stone projecting from a wall to support a weight.

DATUM — An assumed level point used as a reference point for the measurements of levels.

DOUBLE-FACED — A wall having front and rear faces.

DRIP — A groove on the under surface of a coping or sill with stones projecting beyond the face of the coping or sill; designed to prevent water from passing from the projection to the wall. Also known as THROAT.

ENTASIS — Convex swelling in a column.

EXTRADOS — The outside of the curve of an arch.

FACE OF STONE — Exposed surface of the stone; the vertical face in elevation.

FACE RETURNED — The vertical face exposed to the side elevation.

FINIAL — The ornament on an apex or a saddle stone.

FOUNDATION — An underlying natural or prepared base or support.

FOUNDS — Short for "foundations."

GABLE — The vertical wall at the end of a pitched roof.

GROUT — Mortar made into a liquid form for pouring into cavities.

HEAD — A stone across the top of an aperture. Also known as a LINTOL or LINTEL.

HOLLOW BEDDING — Setting sill ends only on mortar, leaving the center part hollow. If the sill were bedded solid, when the load was applied (or with settlement), the sill would crack.

HOOD MOULD — Stones over a door or window opening, to throw off rainwater. Also called LABEL COURSE.

INFILL — Small stones used between faces of a wall.

INTRADOS — The underside of the curve of an arch.

JOGGLE — An indentation cut in the joint surfaces of stone.

JUMPER — A tool used to drill holes in stone.

KNEELER — A long stone with the coping worked in it, tailing it into a gable wall. Serves to resist the sliding tendency of the coping.

LABEL COURSE — See HOOD MOULD.

LIFTS — Height of courses.

LINTOL — A beam across the top of an aperture. (Also LINTEL.)

MORTAR — A mixture of cement and lime with water and sand, which hardens and is used to join bricks or stone. (Also called "mud.")

OFF BED — Stone laid on edge, with its natural bed exposed.

OUT OF TWIST — A stone surface that is true in all ways.

PLUMB — 1. A lead weight attached to a cord and used to indicate a vertical line. 2. Exactly vertical or true.

PLUMBINGS — The corners or first part built, acting as guidelines in keeping a wall vertically true and horizontally level.

QUOIN — A stone laid at the external angle of a building, so it is a header (inband) in respect to the wall proper, and a stretcher (outband) in respect to the return wall. Also called RYBIT.

RANDOM RUBBLE — The early Celtic art of building with irregular stones bedded on mortar.

RISERS — Stones allowed to rise above your line in rubble work, to break the straight line effect.

RYBIT — See QUOIN.

SADDLE STONE — See APEX STONE.

SAVER ARCH — A flat arch consisting usually of three stones with the center stone acting as a keystone.

SCARCEMENT — The ledge formed at a place where part of the wall is set back, especially the footings of a stone wall.

SHAG POINTING — Method of pointing that uses the trowel; used only in random rubble.

SKEW CORBEL — A projecting stone at the lowest point of the triangular portion of the gable end of a wall.

SNOUTS — The farthest projection of a rough stone face, in rubble work.

SILL — Stone forming the lower boundary of a door or window opening.

SPANDRIL — Surface of the main wall contained by the horizontal line from the crown; a vertical line from the lowest point on the extrados, and that portion of the extrados between these two points.

SPRINGER — The lowest point of an arch.

STOOLING — Point on which the mullion, transom, and head rests. A dowel is used in this to prevent any movement. Also STOOL, SEAT, BASE.

STRINGS — String course, a projecting stone.

TEMPLATE — A pattern or mould used as a guide to the form of a piece being made.

THROAT, THROATED — See DRIP.

TIE OVER — To lay the stone so that one stone overlaps the other, locking all together.

TINGLE — To prevent the building line from sagging, a twisted piece of wire holds the line free, with a small stone or brick holding the wire in position.

TRAMMEL — A form of compass used for working out large circles.

TYMPANUM — The space within an arch and above a lintel.

VOUSSOIR — Stone in the course of an arch.

WALL RESTS — The top part of a wall on which the load rests.

WEATHERED — Stone that is worked away to the edge to allow water to get away quickly so it will not collect and soak into the stone.

Index

About the Author

Ian Cramb was born February 20, 1928, the elder son of a local building contractor. He was raised and educated in the small Burgh of Dunblane, Perthshire, Scotland. The Cramb family have been stonemasons since 1770; building contractors since 1864.

Ian Cramb started his apprenticeship on March 1, 1942, the fifth generation to do so. He took evening classes three nights a week, learning building construction, for a period of five years.

Dunblane was an ideal area for an apprentice stonemason, as every building is constructed with stone. Surrounded by large estates, farm buildings, a ruined 13th century bishop's palace, two large 15th century castles, a Gothic cathedral, and numerous other old buildings, it was an apprentice stonemason's paradise.

In 1951, Mr. Cramb took over as stone foreman on a large restoration and rebuilding project in the old "top part" of the town of Stirling. This "top part" consisted of the streets that led up to the highest point in the town, the Castle.

Most of the buildings, built around 1420-60, were being demolished, and new stone buildings built with exteriors conforming to the 1420s style. In others, the exteriors were preserved, with only the interiors modernized.

One of the buildings in this restoration project was the Nursery of King James VI of Scotland, son of Mary Queen of Scots. This complete project took five years. During the course of this project, Mr. Cramb spent his free time and vacations working on projects at the historic Abbey of Iona. At the end of this period he began to specialize in restoration of historic buildings.

In 1955, Mr. Cramb was invited to visit the United States to study the American way of building, but as it turned out, the Americans preferred to learn his methods. In April of 1956 he arrived at Yale University for a six-month stay, which included visits to different sites in the U.S., among them St. Albans Cathedral in Washington, DC as well as many other buildings along the East Coast.

He went back to Iona at the end of August, where he met the Queen and other members of the royal family, for a dedication of the work on the restoration of Iona Abbey.

In January of 1957, Mr. Cramb took over as Master stonemason on the restoration of the monastic buildings around the Abbey. He rebuilt the Cloisters, which consisted of fifty-seven depressed Gothic arches, and hand dressed and built two large semi-circular arches for the corners. He refers to this as "my own Sermon in Stone."

Also on Iona, he dressed stone arches and restored St. Michael's Chapel, and restored St. Oran's Chapel in the Cemetery of Kings, built in 1075. He also restored and preserved a 13th century parish church.

Mr. Cramb and his wife left for Washington, DC in August of 1959. He set stone and marble on the Capitol Building, setting a marble star at the main entrance; he acted as stone and marble mason for the Raeburn Building on Independence Avenue and for the World Bank Building on Pennsylvania Avenue.

He returned to Scotland in 1962 to act first as Inspector of Works, Building, then as Deputy Burgh Engineer for the Burgh of Wick. He then went back to his trade as stonemason in Edinburgh.

In 1984 he took over as master restoration and building stonemason at Mugdock Castle, building of which was begun in the 13th century and added to in the 14th through 17th centuries. Each part of the Castle required different building methods.

He has acted as Architect's Consultant on a number of projects involving castles and other old buildings.

Mr. Cramb has diplomas in building construction, as Inspector of Works, and in architectural drawing for complete buildings (including castles) as well as in stonework. He is an expert on all styles of arch construction. He received the designation of Master at his trade upon completion of the Cloisters at Iona Abbey.

In 1987 Mr. Cramb returned to the U.S. to pass the secrets of his trade on to his eldest son.

His stone-cutting tools and mallet are about 150 years old, having been passed down from father to son. They were given to The Carpenters Guild, Carpenters Hall, Philadelphia, and now can be found in the Historic Building Museum in Washington, DC.